CDM Regulations 2007
Explained

Raymond Joyce BSc, MSc, LLB,
DIC, CEng, FICE, ACIArb
Chartered civil engineer and
solicitor of the Supreme Court

thomas telford

Published by Thomas Telford Publishing, Thomas Telford Ltd, 1 Heron Quay, London E14 4JD. www.thomastelford.com

Distributors for Thomas Telford books are
USA: ASCE Press, 1801 Alexander Bell Drive, Reston, VA 20191-4400, USA
Japan: Maruzen Co. Ltd, Book Department, 310 Nihonbashi 2-chome, Chuo-ku, Tokyo 103
Australia: DA Books and Journals, 648 Whitehorse Road, Mitcham 3132, Victoria

First published 2007

Reprinted 2008

A guide to health and safety prosecutions. G. Forlin and M. Appleby.
ISBN 978-07277-3481-5
Engineer's dispute resolution handbook. R. Gaitskell. ISBN 978-07277-3450-1
Construction Law Handbook 2008 edition. Eds Sir V. Ramsey *et al.*
ISBN 978-07277-3567-6

A catalogue record for this book is available from the British Library

ISBN: 978-0-7277-3496-9

This book is published on the understanding that the author is solely responsible for the statements made and opinions expressed in it and that its publication does not necessarily imply that such statements and/or opinions are or reflect the views or opinions of the publishers. While every effort has been made to ensure that the statements made and the opinions expressed in this publication provide a safe and accurate guide, no liability or responsibility can be accepted in this respect by the author or publishers.

Typeset by Academic + Technical, Bristol
Printed and bound in Great Britain by MPG Books, Bodmin, Cornwall

In my judgment the most striking feature is the clear lack of liaison within and between the defendants in relation to matters of safety. This is in no way inconsistent with the existence of an undoubtedly close working relationship ...

The Abbeystead Disaster
Eckersley & Others v *Binnie & Others*
Rose J

Contents

Preface

Analysing health and safety statistics is a sombre affair. A career in occupational health and safety demands tenacity because the successes are not defined as moments of cheer as compared with the awfulness of failure. The Construction (Design and Management) Regulations 2007 are a moral and practical support to remind all of us that we cannot be complacent about the application of management and technology to reduce the opportunities that exist to maim and kill workers in the construction industry. The 1994 Regulations had their critics but did they really do such a bad job so as to justify a complete reworking to produce the new Regulations?

I welcome the fact that the Construction (Health, Safety and Welfare) Regulations are now incorporated in the new Regulations, thus combining the requirements of the Temporary or Mobile Construction Worksites Directive into one piece of legislation. However, I am concerned that there will be many within the construction industry, and those who require its services, who do not fully appreciate that the new Regulations are significantly different and realign responsibilities and liabilities for health and safety.

I have not embarked upon a comparison between 1994 Regulations and the new Regulations because it is, it appears to me, a pointless exercise when the 1994 Regulations have already been consigned to history. The clients, those who employ the services of the construction industry, have been given a greater responsibility by the new Regulations for the overall management of health and safety on construction sites. This shift of responsibility has been a cause of great concern to various client organisations which culminated in the Leader of the Opposition tabling an Early Day Motion requiring a debate as to whether the Regulations should have been annulled on the basis that they were unfair to clients.

I have spent more than 12 years advising clients on the extent of their duties under the 1994 Regulations. On many occasions I have had to follow this advice with a gentle and diplomatic rebuke that it was not an appropriate approach to see what measures could be taken to avoid the full rigour of the 1994 Regulations. The fact that the new Regulations do not allow clients to appoint agents to undertake the client duties for them is a concerted move to ensure that any parties to a project cannot shirk their health and safety management responsibilities.

This new book follows the format of the previous two editions on the 1994 Regulations in being organised so that chapters emphasise the main legal responsibilities of the named duty holders. This should assist the reader interested in one particular role to quickly assimilate the legal responsibilities attached to that duty holder.

My explanation of the new Regulations is based on my own views and opinions, although, in practice, many of my predictions, especially about an unnecessary bureaucratic burden, came to pass since my first edition on the 1994 Regulations. Regrettably, despite the Executive's intention that the new Regulations will impose less of a bureaucratic burden, I fear that a safety-first approach will be adopted by most parties to a project which will result in no less bureaucracy. The challenge for all parties to a construction project is applying the right amount of judgment, but who can know if the judgment applied by a health and safety inspector or the courts will coincide with the judgment of the parties.

Thanks are due to Shanette Jones, my personal assistant, without whose organisational support the completion of this book could not have been achieved. The support and tolerance I have had from my wife, Yvonne, has made writing this book easier than it might have been and puts me even deeper in her debt.

I am also indebted to the very many organisations and individuals that have sought my advice over the years which has provided me with a deeper insight into the Regulations than I might otherwise have had.

Raymond Joyce

May 2007

Table of Cases

Table of UK Statutes

Table of Statutory Instruments

Table of abbreviations

1994 Regulations	the Construction (Design and Management) Regulations 1994
the **ACOP**	the Approved Code of Practice
the **C(HSW) Regulations**	The Construction (Health, Safety and Welfare) Regulations 1996
the **Commission**	the Health and Safety Commission
the **EU**	the European Union
the **Executive**	the Health and Safety Executive
the **Framework Directive**	Council Directive of 12 June 1989 on the introduction of measures to encourage improvements in the safety and health of workers at work (89/391/EEC)
the **Management Regulations**	the Management of Health and Safety at Work Regulations 1999
HASWA 1974	the Health and Safety at Work etc. Act 1974
the **Regulations**	the Construction (Design and Management) Regulations 2007
the **Temporary or Mobile Construction Sites Directive**	Council Directive of 25 June 1992 on the implementation of minimum safety and health requirements at temporary or mobile construction sites (92/57/EEC)

Related publications and further reading

Health and Safety Executive, *Occupational exposure limited* (revised periodically), EH40, HSE Books, 2005

Construction Industry Research and Information Association, *Construction work sector guidance for designers*, C662, CDM 2007, Ove Arup and Partners, CIRIA Books, 2007

Construction Industry Research and Information Association, *Safe access for maintenance and repair*, C611, Iddon, J. and Carpenter, J, CIRIA Books, 2004

Construction Industry Research and Information Association, *Good practice guidance for refurbishing occupied buildings*, C621, W. Fawcett, Palmer, J., CIRIA Books, 2004

Construction Industry Research and Information Association, *Safer surfaces to walk on: reducing the risk of slipping*, C652, Carpenter, C., Lazarus, J. and Perkins, D., CIRIA Books, 2006

Construction Industry Research and Information Association, *Workplace 'in-use' guidance for designers*, C663 CDM 2007, Gilbertson, A., CIRIA Books, 2007-07-23

Health and Safety Executive, *New CDM Approved Code of Practice*, L144, HSE Books, 2007

Health and Safety Executive, *Health and safety in roof work*, HSG33, HSE Books, 1998

Health and Safety Executive, *Avoiding danger from underground services*, HSG47, HSE Books, 2000

Health and Safety Executive, *The safe use of vehicles on construction sites*, HSG 144, HSE Books, 2000

Health and Safety Executive, *Backs for the future – safe manual handling in construction*, HSG149, HSE Books, 2006

Health and Safety Executive, *Health and safety in construction*, HSG 150, HSE Books, 1997

Health and Safety Executive, *Protecting the public – your next move*, HSG151, HSE Books, 1997

Health and Safety Executive, *Fire safety in construction work*, HSG168, HSG Books, 1997

Health and Safety Executive, *Be safe and shore – H&S in excavations*, HSG185, HSE Books, 1999

Health and Safety Executive, *Introduction to asbestos in workplace buildings*, HSG213, HSE Books 2001

Health and Safety Executive, *Managing asbestos in workplace buildings*, INDG223, HSE Books, 2003

Health and Safety Executive, *Handling heavy blocks*, CIS37, HSE Books, first published 1993, reprinted 2003

Health and Safety Executive, *Noise in construction*, INDG127, HSE Books, 1994

Health and Safety Executive, *Upper limb disorders – assessing the risks*, INDG171, HSE Books, 2005

Health and Safety Executive, *Handling kerbs: reducing the risks of musculoskeletal disorders*, CIS57, HSE Books, 2005

Health and Safety Executive, *The high 5 – five ways to reduce risk on site*, HSE Books, 2003

Health and Safety Executive, *Safe erection, use and dismantling of false work*, CIS56, HSE Books, 2003

Health and Safety Executive, *Cement*, CIS26, revision 2, HSE Books, 2002

Health and Safety Executive, *Safety in excavations*, CIS8 revision 1, HSE Books, 2006

Health and Safety Executive, *Establishing exclusion zones when using explosives in demolition*, CIS45, HSE Books, first published 1995, reprinted 2002

Health and Safety Executive, *Inspections and reports*, CIS47, revision, HSE Books, 2005

Health and Safety Executive, *Construction fire safety*, CIS51, revision 1, HSE Books, first published 1997, reprinted 2005

Health and Safety Executive, *Construction site transport safety: safe use of site dumpers*, CIS52, HSE Books, 2006

Health and Safety Executive, *Crossing high-speed roads on foot during temporary traffic management works*, CIS53, HSE Books, 2000

Health and Safety Executive, *Dust control on concrete saws used in construction industry*, CIS54, HSE Books, 2000

Health and Safety Executive, *The absolutely essential toolkit for the smaller construction contractor*, HSE Books, 2002

Health and Safety Executive, *Portable nuclear moisture/density gauges in the construction industry*, IIS3, HSE Books, 2002

Health and Safety Executive, *Chemical cleaners*, CIS24, revision 1, HSE Books, first published 1998, reprinted 2003

Health and Safety Executive, *Solvents*, CIS27, revision 2, HSE Books, 2003

Health and Safety Executive, *Silica*, CIS36, revision 1, HSE Books, first published 1999, reprinted 2004

Health and Safety Executive, *Portland cement dust, hazard assessment document*, EH75/7, HSE Books, 2006

Health and Safety Executive, *Inspecting fall arrest equipment made from webbing or rope*, HSE Books, first published 2002, reprinted 2006

Health and Safety Executive, *Tower scaffolds*, CIS10, revision 4, HSE Books, 2005

Health and Safety Executive, *Issues surrounding the failure of an energy absorbing lanyard*. Thomas, D., SIR59, HSE Books 2001

Health and Safety Executive, *Preventing falls from boom-type mobile elevating work platforms*, HSE Information Sheet MISC614, HSE Books, first published 2003, reprinted 2006

Health and Safety Executive, *Welfare provision at transient construction sites*, HSE Books, first published 1997, reprinted 2006

Health and Safety *Executive, Provision of welfare facilities at fixed construction sites*, CIS18, revision 1, HSE Books, first published 1998, reprinted 2004

Better Regulations Executive, *Principles of good regulation,* http://www.cabinetoffice.gov.uk/regulation/consultation/consultation guidance/planning a consultation/principles good regulation. asp

1 Introduction

The facts
The analysis
Why the need for the new CDM Regulations?
Cost benefit

The facts

The health and safety record of the construction industry in the 1980s was such that the Health and Safety Commission planned to introduce far-reaching legislation targeted at the management of construction projects. However, these plans were overtaken by the Temporary and Mobile Construction Sites Directive which would achieve a very similar result when implemented into domestic law. Therefore, the 1994 Regulations came into being for the purposes of implementing the Temporary and Mobile Construction Sites Directive but there is no doubt that similar legislation would have come into being in any event.

The statistics produced by the Executive for fatalities and serious injuries in the construction industry have shown a downward trend since 1964. During the 10 years from 1964 to 1974 the number of deaths per year in the construction industry fell from 276 to 166 which was an improvement of 40%. In the following 10 years from 1974 to 1984 the number of deaths per year in the construction industry fell from 166 to 100, a further reduction by 40%.

The statistics for deaths per 100 000 workers are the preferred basis for analysing the statistics because this normalises the overall number of fatalities regardless of the level of activity within the construction industry. Therefore, in the period from 1985 to 1994/5 the improvement in the annual statistics for fatalities was 8.6 to 5.1 deaths per 100 000 workers. This was an improvement of 40% without the benefit of the 1994 Regulations.

The reasons for the improvements in the statistics for fatalities in the construction industry prior to the 1994 Regulations can be attributed to a number of factors:

1. increasing the awareness of the need for improved health and safety management culminating in the Health and Safety at Work etc. Act 1974;

2. regulations made under the Health and Safety at Work etc. Act 1974 including the 'Hard Hat' Regulations;

3. improvements in equipment and machinery; and

4. developments in construction techniques.

Despite the general downward trend of improvement in the health and safety statistics for fatalities, the Executive concluded in *'Blackspot Construction'*[1] that 70% of the deaths could have been prevented by positive action by managers within the construction industry. It was the Executive's opinion that the majority of the accidents could be prevented by the application of reasonable practicable precautions. The analysis of the main causes of accidents by the Executive revealed that most were due to the following reasons:

1. A lack of supervision by line managers in the industry. This was identified by the Executive, and was a view supported by the Trade Unions. It was felt that the widespread use of sub-contractors and self-employed labour led to problems of management and control which were exacerbated by new forms of contracting which involved management remote from the site.

2. Custom and practice in the industry was not assisting workers or equipping them to identify dangers and empowering them to take steps to protect themselves.

3. A lack of co-ordination between the members of the professional team at the pre-construction stage.

The preamble to the Temporary or Mobile Construction Sites Directive on the implementation of minimum health and safety requirements on temporary or mobile construction sites concurred in recognising the importance of the pre-construction phase as a cause of accidents across the European Union as follows:

> *... unsatisfactory and/or organisational options or poor planning of the works at the project operations stage have played a role in more than half of the occupational accidents occurring on construction sites in the Community.*

So the 1994 Regulations came into force on the back of high expectations that they could contribute to a significant improvement in the health and safety management and therefore a reduction of fatalities in the construction industry.

In the 10 years following the introduction of the 1994 Regulations, the improvement in the annual statistics for fatalities was 5 to 3.5 per 100 000 workers, representing an improvement of 30%. There had been some upward blips in the 10 years after the introduction of the 1994 Regulations but the trend was undeniably downward.

The analysis

During the 10 years after the introduction of the 1994 Regulations, the statistics for the number of deaths per 100 000 workers continued the general downward trend in the construction industry; but, were the 1994 Regulations responsible for the continuation of that trend or did they, in fact, make no difference to the statistics at all?

The difficulty with analysing statistics that are either expressed as an absolute number of deaths or a number of deaths per 100 000 workers is the extent to which they can be distorted by one or two significant accidents or incidents within the relevant year and by the complex interdependencies of many disparate contributory factors. Without the 1994 Regulations it is arguable that the construction industry would still have benefited from a continuing improvement in the statistics for fatalities.

In developing estimates for the costs of implementing the 1994 Regulations the Health and Safety Commission recognised that the reduction in the level of accidents would be the principal quantifiable benefit. The assumptions adopted were that accidents would be reduced by 33% on small to medium sites if the 1994 Regulations were implemented. On large sites, where safety management was usually better developed, a 20% reduction in accidents was assumed.

During the first 10 years of the 1994 Regulations the improvement across all sites for fatalities was 30% which, the Commission could argue, was in line with expectations. However, without the 1994 Regulations, improvements would have been likely to come about as a result of the following:

1.　the public's expectation to see improvements in the working environment;

2.　continuing improvements in the construction techniques;

3.　increasing investment by the construction industry in health and safety management; and

4.   compliance with other relevant health and safety regulations.

Thus, it is impossible to say the 1994 Regulations have been a success in improving the health and safety record of the construction industry. Although compliance with the 1994 Regulations will have been a contributory factor, no one can know with any certainty to what extent lives were saved.

The 1994 Regulations did make a difference in one very important regard and that was their contribution to heightening awareness of health and safety in the construction industry and promoting communication between the parties. It is the promotion of improved communication between the parties which, in the opinion of the author, has done as much to reduce the number of contract disputes as the provisions of the Housing Grants, Construction and Regeneration Act 1996.

Why the need for the new CDM Regulations?

The 1994 Regulations were in force for just over 12 years, and in that time the debate about the their effectiveness and the unnecessary bureaucracy was never off the agenda within the construction industry. The constant criticism of the bureaucracy for which the 1994 Regulations were responsible did not persuade any side of the construction industry that the 1994 Regulations were generally of no benefit. Inevitably, in the opinion of many organisations, the Regulations could be improved. The results of reviewing the management of health and safety in practice together with consultation and debate led to the decision by the Commission to revise the 1994 Regulations.

The new Regulations are at the very least a radical revision, insofar as the new Regulations were still constrained within the remit of continuing to implement the Temporary or Mobile Construction Sites Directive. In fact, the new Regulations represent a virtual rewrite with very little of the original text remaining unamended or rewritten.

The Regulations have been developed in line with Better Regulation principles, and are intended to reduce bureaucracy and improve the commercial and business management of construction projects. The Commission's aims for revising the 1994 Regulations were to further reduce construction accidents and ill health by the following means:

1.   being clearer – making it easier for duty holders to know what is
       expected of them;

2. being flexible and accommodating a wide range of contractual arrangements found in the construction industry;

3. emphasising the need to plan and manage work, rather than treating the paperwork as an end in itself;

4. emphasising the communication and co-ordination advantages of duty holders working in integrated teams; and

5. simplifying the way duty holders assess competence.

Despite wide consultation, it is inevitable that the Regulations are not without their detractors. At the eleventh hour, within days of the Regulations coming into force, the Opposition tabled an Early Day Motion in Parliament. The Opposition had been lobbied by the various client organisations on the basis that clients were at an unfair risk of prosecution. The Early Day Motion was a means of requesting a Parliamentary debate that the Regulations should be annulled. The clients' role under the Regulations is undoubtedly more onerous than it had been under the 1994 Regulations. Very few persons would deny that clients have a significant influence over health and safety on a project and for many the additional responsibility was to be welcomed. The debate took place on 10 May 2007 and the Early Day Motion was defeated by the Government thus leaving the Regulations in place and unamended.

The organisations representing clients continue to disagree with the Executive over the content of the industry guidance for the Regulations on the basis that it does not go far enough to protect clients. It was only to be expected that the critics of the 1994 Regulations would regroup and refocus their criticism of the new Regulations.

Finally, there are also those who complain that the attempt to reduce the bureaucracy and the costs to the construction industry will fail because of the increased need to comply with the assessment of competence and the Commission's optimistic assumptions as to improvements in the future health and safety statistics.

Cost benefit

When the 1994 Regulations were introduced, the Executive acknowledged that there would be a significant cost associated with improving health and safety management in organisations that did not already have a well-developed approach. In the consultative document for the 1994 Regulations, the Commission assumed (while acknowledging the difficulties) that the total cost to the construction industry in implementing the 1994 Regulations would be in the region of £550 million, based on an industry output of £37 billion for 1991. The two main areas of additional costs were associated with the duties upon designers and the duty upon the now revoked role of planning supervisor and principal contractor to produce a health and safety plan. The Commission estimated that compliance by the designers with their new duties might cost up to an additional £292 million each year, and by the planning supervisor and principal contractor an additional £185 million each year.

In arriving at their estimates, the Commission used the assumptions for the improvement in the level of accidents as the principal quantifying benefit.

There has been no definitive analysis of how the costs of implementing the 1994 Regulations have been incurred and how they compare with the original estimates. Owing to the lack of resources to collect any organised data from the huge number of construction projects it is virtually impossible to make any assessment of the savings to the construction industry, if any, by improving quality and efficiency as a result of improvements in health and safety management.

It is claimed by the Commission that the new Regulations will save the industry £2.7 billion, through reduced bureaucracy and a further benefit to the construction industry by up to £3.2 billion by reduced loss of working time. However, on figures representing a worst case, the Commission has conceded that the Regulations could cost the construction industry £660 million.

The cost benefit of the new Regulations has been criticised on the basis that the estimates are based on the benefit arising from a 50% reduction in the number of deaths and injuries. The Commission has calculated that the new Regulations will also reduce construction costs by 3% and drastically reduce the cost of competence assessments for the various duty holders. This is highly optimistic since the core criteria for assessing competence as set out in the Approved Code of Practice are likely to increase the cost of competence assessments.

A debate about the cost benefit and the assumptions adopted by the Health and Safety Commission is in reality a debate without end and without any benefit of its own. The actual outturn costs will never be identified and the debate is highly unlikely to contribute to any future amendments to the Regulations.

The construction industry has no option other than to implement the Regulations. It will inevitably bring about higher costs associated with training and changes to procedures at the outset. Perhaps those costs are a reasonable price to pay to boost the improvement in health and safety management. The Regulations cannot protect workers from their own negligence or inadvertence but if duty holders focus on the identification of hazards and risks and apply appropriate judgement based on their own experience and competency the public at large can expect a continuing improvement and downward trend in the number of deaths and accidents in the construction industry.

Note

1. *Blackspot Construction* published by the Health and Safety Executive 1988.

2 Framework of health and safety law

Introduction
Common law
   Reasonably practicable
UK legislation
European health and safety initiative

Introduction

The Regulations are a small part of a very much larger legal framework of law in Great Britain and the European Union which addresses matters of health and safety.

The Regulations, except for some minor repeals, revocations and modifications set out in regulation 48, do not change the general principles of health and safety law. They are, however, a re-enactment of the European Union's Temporary or Mobile Construction Sites Directive and a continuing development of the widely accepted need to influence and regulate management practices to improve health and safety in the construction industry.

Common law

Common law is often referred to as that part of English law which cannot be found in Acts of Parliament but in the decisions of the courts and custom. Common law principles change and develop with the decisions of the courts. Conversely, Acts of Parliament can abolish well-established rules of common law.

The importance of common law in health and safety matters has diminished considerably with the increased amount of legislation. Statute law will always prevail whenever there is a conflict with the common law, thus upholding the legislative supremacy of Parliament. However, despite the extensive statutory legislation which affects matters of health and safety, common law principles co-exist with statutory provisions. For example, there is implied into every contract of employment a term that the employer will provide safe plant and premises, a safe system of work and reasonably competent fellow workers.

The common law, insofar as it is based on decisions of the courts, also serves to give effect to the intentions of Parliament in the interpretation of Statutes.

Reasonably practicable

Where an employer's obligation is qualified by the test of SFAIRP, the employer can escape liability if it can show it has done everything reasonably possible to avoid risks to the health and safety of workers. The leading case which established the guiding principle as to the interpretation of 'reasonably practicable' is *Edwards* v. *National Coal Board* [1949] All ER 743(CA). Asquith LJ stated as follows:

> *'Reasonably practicable' is a narrower term than 'physically possible' and seems to me to imply that a computation must be made by the owner in which the quantum of risk is placed on one scale and the sacrifice involved in the measures necessary for averting the risk (whether in money, time or trouble) is placed in the other, and that, if it is shown that there is a gross disproportion between them – the risk being insignificant in relation to the sacrifice – the defence has discharged the onus on them. Moreover, this computation falls to be made by the owner at a point in time and anterior to the accident. The questions he has to answer are:*
>
> *(a) What measures are necessary and sufficient to prevent any breach ...?; and*
>
> *(b) are these measures reasonably practicable?*

Therefore, to determine what is reasonably practicable, a cost benefit exercise has to be carried out. If the risk is insignificant in relation to the sacrifice it is not reasonably practicable to take steps necessary to control a hazard.

Based upon the above judgment, *Croner's Health & Safety at Work* (1994) gives the following definition of reasonably practicable:

> *The duty to do what is reasonably practicable is less strict than the unqualified work 'reasonably' implies that a computation must be made in which the question of risk is placed on one side and the sacrifice involved in instituting the measures necessary for alleviating the risk (whether in money,*

time or trouble) is placed on the other. If, when this 'cost benefit' exercise has been carried out, the risk is insignificant in relation to the sacrifice, it is not reasonably practicable to take the steps necessary to control the hazard.

Thus, an employer or duty holder who conducts a cost benefit analysis and concludes that the costs of implementing health and safety procedures outweigh the benefit is, in theory, not obliged to take any further action. But, a word of warning, as a basis for a defence such arguments are not convincing except in the most compelling circumstances.

Prior to June 2007 the European Commission had argued that the Framework Directive should impose strict liability on employers for any circumstances that are a risk to the health and safety of its works, thus depriving an employer of a defence that the costs of prevention would have been grossly disproportionate.

The only exceptions to escape strict liability in the European Commission's view were 'unusual or unforeseeable circumstances, beyond the employer's control or to exception events, the consequences of which could not have been avoided despite the exercise of all due care' (Article 5(4) of the Framework Directive).

The expression 'so far as is reasonable practicable' ('SFAIRP') was challenged by the European Commission on the basis that it qualifies UK employers' obligations in a way that is inconsistent with the Framework Directive. Fortunately the European Court of Justice ruled in June 2007 that SFAIRP is not unlawful, meaning measures to reduce risks do not have to be grossly disproportionate.

UK legislation

Numerous Acts of Parliament and statutory instruments have come into force with the intention of regulating the improving health and safety management in the construction industry. Of prime importance is the Health and Safety at Work etc. Act 1974.

HASWA 1974 established the Health and Safety Commission ('the Commission') and gave it the power to propose health and safety regulations and approve codes of practice. It also set up the Health and Safety Executive with responsibility for enforcing health and safety laws.

The radical difference between HASWA 1974 and all preceding health and safety legislation was the emphasis on individuals and their duties as compared with premises. Rather than a prescriptive approach, HASWA 1974 is based on principles designed to bring about a greater awareness of the problems associated with health and safety issues. The legislation is addressed to individuals and includes employers, employees and producers of industrial products.

This is the primary safety legislation in the UK, HASWA 1974 being the Act under which virtually all subsequent health and safety regulations have been made. All the regulations which implement the EU Directives referred to below were made under the powers granted by HASWA 1974.

The Management Regulations are one of the most important regulations, which cover all areas of work, except sea transport, and apply to all businesses regardless of size. However, the nature, extent and cost of the measures that need to be taken to ensure compliance will vary, depending on the nature of the hazards and level of risks associated with different industries.

The three main requirements of the Management Regulations concern:

1. risk assessment;

2. arrangements for protective and preventative measures; and

3. the appointment of competent persons to assist with protective and preventative measures.

They require an employer and self-employed employer to implement any preventative and protective measures on the basis of the following principles, set out at Schedule 1 of the Management Regulations which apply equally to construction sites:

(a) avoiding risks;

(b) evaluating the risks that cannot be avoided;

(c) combating the risks at source;

(d) adapting the work to the individual, especially as regards the design of workplaces, the choice of work equipment and the choice of working and production methods, with a view, in particular, to alleviating monotonous work and work at a predetermined work rate and to reducing their effect on health;

(e) adapting to technical progress;

(f) replacing the dangerous by the non-dangerous or the less dangerous;

(g) developing a coherent overall prevention policy that covers technology, organisation of work, working conditions, social relationships and the influence of factors relating to the working environment;

(h)   giving collective protective measures priority over individual protective measures; and

(i)   giving appropriate instructions to employees.

European health and safety initiative

The European Communities Act 1972, which came into effect from 1 January 1973, incorporated the Treaties constituting the European Economic Community, now called the European Union (EU), into United Kingdom legislation. EU law, which includes Directives, is implemented as a domestic law in the United Kingdom by legislation passed or approved by Parliament either by Acts or by Statutory Instruments.

The Single European Act introduced Article 118A into the Treaty of Rome which obliges the Member States to 'pay particular attention to encouraging improvements, especially in the working environment, as regards the health and safety of workers'. Article 118A also provides that the health and safety policy would be introduced, by the adoption of Directives on working conditions and technical standards, on the basis of majority voting in the EU Council of Ministers. The Directives adopted under Article 118A are intended to avoid, as far as possible, imposing administrative, financial and legal constraints that would 'hold back the creation and development of small and medium-sized undertakings'. The Directives are only intended to establish minimum standards. Member States are not prevented from maintaining or introducing more stringent measures, providing the measures do not interfere with other objectives of the Treaty of Rome, including measures which would have an anti-competitive influence on the open market. To a greater or lesser extent, all the Directives adopted under Article 118A and the regulations which implement them, have an impact on the construction industry, and share the same overall strategic objective with the Regulations which form the subject matter of this book: to improve the health and safety of workers.

The Directives which have been adopted under Article 118A include the following:

(i)   Council Directive of 12 June 1989 on the introduction of measures to encourage improvements in the health and safety of workers at work (89/391/EEC) (the Framework Directive). The Framework

Directive is so called because it created the framework for further directives. By creating broad and general duties on employers, employees and the self-employed, the need for dealing with specific hazards and work situations was dealt with in subsequent Directives. This directive has been implemented by the Management of Health and Safety at Work Regulations 1999.

(ii) Council Directive of 30 November 1989 concerning the minimum safety and health requirements for the workplace (first individual directive within the meaning of Article 16(1) of Directive 89/391/ EEC)(89/654/EEC). This directive has been implemented by the Workplace (Health, Safety and Welfare) Regulations 1992 (as amended).

(iii) Council Directive of 30 November 1989 concerning the minimum safety and health requirements for the use of work equipment by workers at work (second individual directive within the meaning of Article 16(1) of Directive 89/391/EEC)(89/655/EEC). This directive has been implemented as the Provision and Use of Work Equipment Regulations 1998 (as amended).

(iv) Council Directive of 30 November 1989 on the minimum health and safety requirements for the use by workers of personal protective equipment at the work place (third individual directive within the meaning of Article 16(1) of Directive 89/391/EEC)(89/656/ EEC). This directive has been implemented as the Personal Protective Equipment at Work Regulations 2002.

(v) Council Directive of 29 May 1990 on the minimum health and safety requirements for the manual handling of loads where there is a risk particularly of back injury to workers (fourth individual directive within the meaning of Article 16(1) of Directive 89/391/ EEC)(90/269/EEC). This directive has been implemented as the Manual Handling Operations Regulations 1992 (as amended).

(vi) Council Directive of 29 May 1990 on the minimum safety and health requirements for work with display screen equipment (fifth individual directive within the meaning of Article 16(1) of Directive 89/391/EEC)(90/270/EEC). This directive has been implemented as the Health and Safety (Display Screen Equipment) Regulations 1992 (as amended).

(vii) Council Directive of 24 June 1992 on the implementation of minimum safety and health requirements at temporary or mobile construction sites (eighth individual directive within the meaning of Article 16(1) of Directive 89/391/EEC)(92/57/EEC). This is the directive which has now been implemented by the Construction (Design and Management) Regulations 2007.

The regulations referred to in paragraphs (ii), (iii), (v) and (vi) above were amended by the Health and Safety (Miscellaneous Amendments) Regulations 2002.

3   An overview of the Regulations

Introduction

The Regulations give effect to Council Directive 92/57/EEC on the implementation of minimum safety and health requirements at temporary or mobile construction sites. Regulation 1 provides:

> *These Regulations may be cited as the Construction (Design and Management) Regulations 2007 and shall come into force on 6th April 2007.*

The Regulations are organised into five parts. All the parts apply to all projects except part 3 which sets out additional management duties for projects that are required to be notified to the Executive. The threshold for notification has not changed from the 1994 Regulations but crucially demolition activity does not make a project notifiable if it would not otherwise be notifiable.

Application of the Regulations

The Regulations apply to England, Scotland and Wales as provided for in regulation 3(1):

These Regulations shall apply –

(a) in Great Britain; and

(b) outside Great Britain as sections 1 to 59 and 80 to 82 of the 1974 act apply by virtue of article 8(1)(a) of the Health and Safety at Work etc. Act 1974 (Application outside Great Britain) order 2001.

Article 8(1)(a) of the Health and Safety at Work etc. Act 1974 (Application outside Great Britain) Order 2001 applies to the following activities or preparation for such activities within territorial waters: construction, reconstruction, alteration, repair, maintenance, cleaning, demolition and dismantling of any building or other structure not being a vessel.

Construction work

The Regulations apply throughout England, Scotland and Wales to all non-domestic construction work. There is no *de minimus* level or intensity of construction activity such that the Regulations would not apply thus regulation 3(2) provides:

'Subject to the following paragraphs of this regulation, these Regulations shall apply to and in relation to construction work.'

It is vital to appreciate the definition of construction work, because any activity which is not within the definition of 'construction work' is not subject to the Regulations.

Construction work in Regulation 2(1) is defined as meaning:

the carrying out of any building, civil engineering or engineering construction work and includes –

(a) the construction, alteration, conversion, fitting out, commissioning, renovation, repair, upkeep, redecoration or other maintenance (including cleaning which involves the use of water or an abrasive at high pressure or the use of corrosive or toxic substances), decommissioning, demolition or dismantling of a structure;

(b) the preparation for an intended structure including site clearance, exploration, investigation (but not site survey) and excavation,

and the clearance or preparation of the site or structure for use or occupation at its conclusion;

(c) the assembly on site of prefabricated elements to form a structure or the disassembly on site of prefabricated elements which, immediately before such disassembly, formed a structure;

(d) the removal of a structure or of any product or waste resulting from demolition or dismantling of a structure or from disassembly of pre-fabricated elements which immediately before such disassembly, formed a structure; and

(e) the installation, commissioning, maintenance, repair or removal of mechanical, electrical, gas, compressed air, hydraulic, telecommunications, computer or similar services which are normally fixed within or to a structure.

The degree of fixity is not described nor elaborated upon in the Approved Code of Practice (ACOP). It is suggested that actual connection is required, as opposed to a free-standing item, e.g. a generator 'fixed' by its own weight

but does not include the exploration for or extraction of mineral resources or activities preparatory thereto carried out at a place where such exploration or extraction is carried out.

Construction work, as defined, associated with the exploration for or extraction of mineral resources is excluded specifically, as is the setting up of any infrastructure for such exploration or extraction. Thus, all work concerned directly with mineral exploration or extraction including deep and opencast coalmining, clay pits, sand, stone and aggregate extraction, is expressly excluded. Only construction work and structures directly concerned with exploration or extraction are exempted. Thus, the Regulations do apply to building, civil engineering or engineering construction work at quarries and mines associated with the development of a site prior to operations to extract mineral resources. They also apply to operations or mines and quarries that are not directly related to the work to extract mineral resources. For clarification purposes, the ACOP lists a number of other activities which do not come within the definition of construction work. The straightforward activities which can be categorised without deep reflection include:

1. Putting up and taking down fabric structures including marquees and tents.

2.   Soft landscaping comprising planting and general horticultural work.

3.   Erection, reposition and moving lightweight panels forming office dividers or stands and displays at exhibitions.

4.   Topographical surveys and examination of structures.

5.   Work to or on marine vessels including ships and mobile offshore drilling rigs.

6.   Offsite manufacture or prefabrication of products for use in construction work including concrete panels and roof trusses.

7.   Manufacture or fabrication of components which will form part of offshore installations and the construction of fixed oil and gas offshore installations.

General maintenance of fixed plant will often be excluded from construction work except where it involves erection of temporary structures for access, dismantling or modification to plant that is more than would be expected for maintenance. This is a matter for consideration and exercising judgement. Key factors would be indicated by an appropriate risk assessment including, for example, the need for ancillary plant, lifting equipment and the time allocated for the task.

The definition of construction work cannot be appreciated fully without a knowledge of the term 'structure', which is defined in regulation 2(1) as:

> *(a) any building, timber, masonry, metal or reinforced concrete structure, railway line or siding, tramway line, dock, harbour, inland navigation, tunnel, shaft, bridge, viaduct, waterworks, reservoir, pipe or pipe-line, cable, aqueduct, sewer, sewage works, gasholder, road, airfield, sea defence works, river works, drainage works, earthworks, lagoon, dam, wall, caisson, mast, tower, pylon, underground tank, earth retaining structure, or structure designed to preserve or alter any natural feature, fixed plant and any other structure similar to the foregoing; or*

> *(b) any formwork, falsework, scaffold or other structure designed or used to provide support or means of access during construction work;*

> *and any reference to a structure includes a part of a structure.*

The types of structure described in (b) are only those used in connection with construction work and specifically excludes grandstands and other temporary platform arrangements used by the entertainment industry.

Notification

Non-notifiable projects

The Regulations apply to all projects, although to a differing extent depending on whether a project is non-notifiable (i.e. small projects) or notifiable. Thus, Part 2 comprising regulations 4 to 13 inclusive will apply to all construction work even when it does not come within the criteria for notification to the Executive, whereas Part 3 applies in respect of notifiable projects.

Despite the declared intention of the Commission to reduce the administrative burden on the construction industry in complying with the 1994 Regulations, the effect of removing the *de minimus* level means that the full effect of Part 2 has to be invoked for every construction project. Thus, regulations 4 to 13 inclusive will apply to the smallest of projects comprising construction work. Controversially, demolition work no longer has to be notified to the Executive, as it had under the 1994 Regulations, thus reinforcing the responsibility on the client when procuring demolition under the non-notifiable project provision.

The simplification of the criteria for triggering a notification may reduce the administrative burden on the Executive but the removal of the *de minimus* level will increase significantly the administrative burden for health and safety on the construction industry.

The Regulations incorporate the 1996 Construction (Health, Safety and Welfare at Work) Regulations, which comprise Part 4 of the Regulations, as confirmed by regulation 3(4), that:

> *Part 4 shall apply only in relation to a construction site.*

The full requirements of the welfare provisions apply to all construction work by virtue of regulation 3(5) which states:

> *Regulations 9(1)(b), 13(7), 22(1)(c), and Schedule 2 shall apply only in relation to persons at work who are carrying out construction work.*

This is as one would expect because a person's welfare needs are not diminished simply because the scale of the construction is not sufficient to be notifiable.

Notifiable projects

Paragraph 3 of regulation 3 states:

> *The duties under Part 3 shall apply only where a project –*
>
> *(a) is notifiable; and*

(b) is carried out for or on behalf of, or by, a client.

The duties under Part 3 are in addition to those in Part 2 and involve an additional duty holder in the guise of the CDM co-ordinator.

The definition of a project in regulation 2(1) means:

> *a project which includes or is intended to include construction work and includes all planning, design, management, or other work involved in a project until the end of the construction phase.*

To ascertain whether a project is notifiable, regulation 2(3) provides that:

> *For the purposes of these Regulations, a project is notifiable if the construction phase is likely to involve more than –*
>
> *(a) 30 days; or*
>
> *(b) 500 person days,*
>
> *of construction work.*

Therefore, a project may involve many months of planning and design but if the construction phase is less than 30 days or involves less than 500 person days the project is not notifiable.

To determine whether Part 3 applies, which includes regulations 14 to 24 inclusive, a client has to decide whether or not the project is notifiable before being committed to appointing a CDM co-ordinator and a principal contractor in accordance with regulation 14. For projects where the duration of the construction work is on the cusp of either of the criteria, a client could hardly be criticised for not notifying a project if, in his considered judgement, the construction phase was likely to involve less than 30 days or less than 500 person days. Unfortunately there is no guidance as to the appropriate test to be applied to the evaluation of 'likely'. On the one hand, there is the ever-present question of acceptable levels of profitability, since there must always be a likelihood that a project will exceed 30 days or 500 person days and, on the other hand, as to whether the test of 'likely' is subjective or objective.

The prudent client should only make the decision as to notification after due consideration and compliance with regulations 4 to 13 and consulting a relevant designer.

The ACOP confirms the approach to counting days by clarifying that holidays and weekends can be discounted if no construction work takes place on those days. This is an extremely important clarification because the financial pressures on a client are such that avoiding the additional administrative burden of Part 3 and the direct costs associated with complying with the additional requirements and obligations will weigh in the decision making whether to notify the Executive. Moreover, is the balance in a client's analysis likely to be affected by the knowledge that the Commission and the Executive have sought to reduce the number of notifications and thereby reduce the Executive's own administrative burden?

Consider the following scenario. A client has a period of 28 days from evacuation of a building in which to undertake construction work including demolition in a building before it is scheduled to be reoccupied. Assuming the construction programme provides for a 7-day working week with a planned construction phase lasting 25 days, to give 3 days 'float' based on an average of 18 persons on site each day the project is not notifiable. But how likely is it that after 450 man days and 25 days of construction the project will be complete? What is the likelihood of an overrun? It is tempting to advise clients to notify whenever in doubt but in so doing the project immediately attracts further costs.

Many projects are of the size in the scenario and under similar constraints and suffer from 'scope creep'. In view of the lack of guidance from the Executive it is unfortunate that clients may be in breach of a fundamental obligation before construction work starts. If the construction work on a project takes longer than had been planned originally, the ACOP confirms that, for a short extension of time or short-term increase in the number of persons, there is no need to notify the Executive. A client should notify the Executive of the project before it is completed if the extension of time or increase in number of persons is significant.

Duty holders

The Regulations bring health and safety management, on an obligatory basis, into the planning and design of construction work, of all projects. No projects are exempted although the size of a project will determine whether it will be notifiable to the Executive. The Regulations direct how the parties to a project will contribute to health and safety management from the outset of planning a project through to completion of construction. The key roles in a project are duty holders and are named as follows.

The client

Only domestic householders, acting as clients, procuring construction work carried out on their domestic premises are exempted from the Regulations.

The client has an overriding duty for any project to ensure that arrangements made for managing the project will be carried out, so far as reasonably practicable, without risk to the health and safety of any person. Ancillary to the construction work the client has also to ensure that there are suitable welfare arrangements for the workers and that a structure to be used as a workplace after construction complies with the Workplace (Health, Safety and Welfare) Regulations 1992.

All projects will need a designer, or designers, to a greater or lesser extent even where a project is not notifiable. However, while it is not a primary obligation for a client to appoint a designer, a client must ensure that a designer and every contractor is given all the relevant information in the client's possession likely to have a bearing on health and safety.

The client has an obligation to appoint a CDM co-ordinator and principal contractor for all projects that are notifiable. The client must be satisfied that all duty holders he appoints to any of the roles are competent and that they have allocated or will allocate sufficient resources, including time, to the project.

The client can no longer appoint an agent, as was permissible under the 1994 Regulations, but is able to elect in writing, where there is more than one client in relation to a project, which client is to be subject to the Regulations.

The CDM co-ordinator

One of the significant changes introduced by the Regulations is the redefining of the role undertaken by the planning supervisor under the 1994 Regulations, which had become overly bureaucratic. Adopting the terminology of the Temporary or Mobile Construction Sites Directive, the modified role is fulfilled by a 'CDM co-ordinator' who has a wide range of duties which, in actual practice, more accurately reflect the roles of designers and contractors.

A CDM co-ordinator is only appointed by a client on notifiable projects and once appointed must advise and assist the client in complying with the Regulations. The CDM co-ordinator has to ensure that arrangements are in place for planning and co-ordinating health and safety measures before and during construction. Although the CDM co-ordinator does not have the obligation to prepare the construction phase plan, other than liaising with the principal contractor, the CDM co-ordinator is required to prepare the health and safety file.

In many ways the role of CDM co-ordinator is more demanding than the former role of planning supervisor owing to the necessity to understand design principles, construction techniques and the continuing duties during the construction phase.

The designer

The designer has a set of duties that apply to all construction work. The influential role of the designer on health and safety management in the construction industry is underscored by the obligation to satisfy him or herself that the client is aware of his or her duties.

The Regulations require the designer to have regard to a list of specific risks which covers the full life cycle of a project including building and operational use thus covering initial construction, maintenance, cleaning, risks in use to persons using the structure as a workplace and finally the demolition which is the final stage of construction work.

In the case of a notifiable project, the designer should not commence work beyond the initial design stage without the confirmation that a CDM co-ordinator has been appointed.

The need to share design information with others, to enable them to comply with the Regulations, is a fundamental requirement. In the case of notifiable projects, the designer is under a duty to ensure that all information that is likely to be needed in the health and safety file is provided to the CDM co-ordinator.

The principal contractor

The role of the principal contractor only exists in respect of notifiable projects where the obligation to prepare a construction phase plan falls upon the principal contractor. The principal contractor has the overall responsibility to plan, manage and monitor the construction phase, so that the construction work is carried out without risks to health or safety. To assist the principal contractor in undertaking its main task, the Regulations empower and require the principal contractor to manage the activities on site by means of disseminating prescribed information, consulting and giving directions to other contractors.

Other contractors

Contractors are required to co-operate with the principal contractor. This includes complying with directions given by the principal contractor and providing him or her with details on the management and prevention of health and safety risks created by the contractors' work on site. Contractors contribute to the management of health and safety on site by the provision of other information to the principal contractor and employees.

Construction phase plan

A co-ordinated approach to the health and safety management of the construction project relies upon communication between the parties. The health and safety plan, which was an innovative feature of the 1994 Regulations, is now reincarnated as the construction phase plan. The construction phase plan links the participants together, with the objective of improving the exchange and communication of document information that affects health and safety.

During the pre-construction phase, the construction phase plan is prepared on information obtained from the CDM co-ordinator, designers and other contractors. Before the construction phase can begin, the construction phase plan must be prepared by the principal contractor and approved by the client. He does this by including details on the management and prevention of health and safety risks created by contractors and sub-contractors. The construction phase plan is a dynamic document subject to continuous review and amendment, fulfilling its role as a co-ordinating mechanism and encouraging co-operation, as construction progresses.

Health and safety file

On the completion of a construction project, the CDM co-ordinator is required to prepare a health and safety file that is handed over to the client. The health and safety file is an important record document and should be easily available to others responsible for later construction work associated with the structure or its maintenance, repair, renovation or demolition.

Revocations and amendments

The Regulations do not change the Health and Safety at Work etc. Act 1974 (HASWA 1974) or repeal any part of it. However, the Regulations either revoke or amend a number of related statutory instruments. The instruments that are revoked or amended are dealt with in Schedules 4 and 5 respectively in regulation 48. The Construction (General Provisions) Regulations 1961 are finally revoked in their entirety and since the Regulations replace the Construction (Design and Management) Regulations 1994, the Construction (Design and Management) (Amendment) Regulations 2000 and the Construction (Health, Safety and Welfare) Regulations 1996, they also are revoked in their entirety.

The Approved Code of Practice and industry guidance

The ACOP, approved by the Health and Safety Commission, with the consent of the Secretary of State for Employment under section 16(1) of the HASWA 1974, provides practical guidance on compliance with the Regulations. Failure to comply with the ACOP is not in itself an offence, although such failure may be taken by a court in criminal proceedings as proof that a person has contravened the Regulations. In such circumstances, however, it is open to a person to satisfy the court that he has complied with the Regulations in some other way.

The industry guidance is drafted by representatives of the design and contracting sides of the construction industry with the intention of providing examples and principles that will assist the relevant duty holders in complying with the Regulations. Industry guidance does not have the status of the ACOP for the purposes of providing a defence. Paradoxically, the failure to comply with industry guidance is more likely to result in an increased risk of civil proceedings for negligence.

4 General principles

Introduction
Co-operation
Co-ordination
Prevention
Writing and documents

Introduction

The introduction of the new duties under the Regulations to 'co-operate' and to 'co-ordinate', at regulations 5 and 6 respectively, can only be observed and adhered to if the parties to a project communicate with each other. It is ironic that, despite the importance of communication between the parties, the word 'communication' is not used in the Regulations. Failures in communication will create or exacerbate risks and more certainly undermine the considerable efforts in planning and managing construction work than any other aspect of planning and management. Compliance with the Regulations can only happen with effective communication and while the duty to communicate is not expressed as a general principle it runs throughout the Regulations as a golden thread with concomitant duties to consult, provide information, display notices, give directions, report, etc.

Co-operation

The duty to co-operate is imposed on every person concerned in a project as set out in regulation 5(1) as follows:

> *Every person concerned in a project on whom a duty is placed by these Regulations, including paragraph (2) shall ...*

Thus, the duty to co-operate is imposed upon the client, designer, contractors, principal contractor and CDM co-ordinator. In addition, other persons working under the control of any of those persons are also subject to the duty as referred to in regulation 5(2) as follows:

> *Every person concerned in a project who is working under the control of another person shall report to that person anything which he is aware is likely to endanger health or safety of himself or others.*

Therefore, not only are persons who are directly involved in construction work under a duty to co-operate, regulation 5(2) imposes the duty to co-operate upon persons 'concerned in the project' who are working under the control of any of the named persons in the Regulations. Such persons, including employees and self-employed, might include administrators, suppliers, off-site inspectors, quality assurance auditors, information technology (IT) suppliers and so on and so on. By this means, the Regulations impose a duty on anybody who might conceivably be involved with a project who is working under the control of one of the named persons. This raises a particular challenge. How can such persons with diverse roles and contributions all be made aware of the Regulations? Bringing to the attention of such persons the duty with which they are fixed is a significant challenge for the named persons under the Regulations.

The ACOP concentrates on the duties of contractors and the principal contractor, with particular regard to site induction and the provision of information, but this overlooks the duty on designers to bring to the attention of their staff, sub-contracted designers and site investigation contractors, their duty to communicate information about the design, choice of materials or circumstances on site which may endanger the health and safety of others.

Clients also have a duty to communicate and consult with their employees and suppliers as part of the information gathering at the outset of a project. In turn the employees and suppliers have a duty to bring to the client's attention any circumstances which they know may endanger the health and safety of others.

There are two sides to the duty of co-operation. Both sides of the duty are mandatory. The first duty is to seek the co-operation of others and the second is to co-operate with those who need your co-operation.

Regulation 5(1)(a) sets out the duty to seek the co-operation of others as follows:

> *Every person concerned in a project on whom a duty is placed by these Regulations including paragraph (2), shall seek the co-operation of any other person concerned in any project involving construction work at the same or an adjoining site so far as is necessary to enable himself to perform any duty or function under these Regulations.*

The obligation to seek the co-operation of others deprives such a person of any defence that they were unable to comply with a duty or function because they were unaware of another person with the appropriate information on the same or adjoining site which was necessary to fulfil that duty or function. The positive obligation means that a person has to make enquiries until such time as they have identified that person who could then provide them with the necessary information.

Conversely, regulation 5(1)(b) imposes an obligation on persons to co-operate as follows:

> *Every person concerned in a project on whom a duty is placed by these Regulations including paragraph (2), shall co-operate with any other person concerned in any project involving construction work at the same or adjoining site so far as is necessary to enable that person to perform any duty or function under these Regulations.*

The duty to co-operate is without any qualification when approached by a person who is fulfilling their duty under regulation 5(1)(a). The person seeking co-operation does not have to be involved in the same project as the person providing the co-operation. Such a person may be involved in a different project on the same site or an adjoining site. It is also possible that the person seeking co-operation of the other may be involved in construction work that is neither concurrent nor sequential to the construction work in which the other person is either involved or has been involved. Thus, the duty to co-operate can apply to a person, who is under a duty by virtue of regulation 5(1)(b), but has no active involvement in the site or an adjoining site.

The duty to seek co-operation is expressed as being to the extent necessary to enable the duties or functions to be fulfilled. The duty is not subject to reasonableness; in other words, the effort needed to comply is whatever it takes to facilitate the performance of any duty or function under the Regulations.

Co-operation can be encouraged by regular multi-disciplinary and multi-organisational meetings and establishing lines of communication that are seen to be effective. The problems associated with the more rigid contractual lines of communication are a real threat to co-operation if a 'claims climate' is prevalent. The contribution by the client pursuant to regulation 9(1) to have ensured that there is sufficient time and resources allocated to the project, can greatly enhance the spirit of co-operation on a project.

Co-ordination

The co-ordination of activities on a project is an essential part of the planning process for the construction phase. This is especially true between the activities of contractors executing construction work. The Regulations impose a general duty of co-ordination on all persons by virtue of regulation 6 which provides:

> *All persons concerned in a project on whom duty is placed by these Regulations shall co-ordinate their activities with one another in a manner which ensures, so far as is reasonably practicable, the health and safety of the persons –*
>
> *(a) carrying out the construction work; and*
>
> *(b) affected by the construction work.*

Other than in emergencies, all persons involved in a project should, as a matter of sound management practice, be aware of the activities of others which will, or will be likely to have an effect on their own activities. Expressed in less than precise terms, the duty under regulation 6 on all persons to co-ordinate their activities is one of the fundamental principles upon which health and safety management can be improved. The importance of co-ordination is highlighted by the ACOP by stating that 'something closer to the construction phase plan will be needed' even for non-notifiable projects where the risks are considered to be high such as structural alterations and deep excavations.

Questions of proportionality and the identity of the person with the responsibility to take the initiative to prepare something less than a construction phase plan create a worryingly vague framework for co-ordination on non-notifiable projects.

Prevention

The general principles of prevention are defined in regulation 2(1) as meaning:

> *the general principles of prevention specified in Schedule 1 to the Management of Health and Safety at Work Regulations 1999.*

Schedule 1 of the Management Regulations lists the principles of prevention as follows:

> *(a) avoiding risks;*

(b) evaluating the risks which cannot be avoided;

(c) combating the risks at source;

(d) adapting the work to the individual, especially as regards the design of workplaces, the choice of work equipment and the choice of working and production methods, with a view, in particular, to alleviating monotonous work and work at a predetermined work-rate and to reducing their effect on health;

(e) adapting to technical progress;

(f) replacing the dangerous by the non-dangerous or the less dangerous;

(g) developing a coherent overall prevention policy which covers technology, organisation or work, working conditions, social relationships and the influence of factors relating to the working environment;

(h) giving collective protective measures priority over individual protective measures; and

(i) giving appropriate instructions to employees.

The importance of the Management Regulations and the general principles of prevention are imported into the Regulations and reinforced by regulation 7(1) which provides:

> Every person on whom a duty is placed by these Regulations in relation to the design, planning and preparation of a project shall take account of the general principles of prevention in the performance of those duties during all the stages of the project.

The scope of regulation 7(1) excludes the construction phase but nonetheless will involve contractors who are involved in the planning and preparation prior to the construction phase. Undoubtedly, this particular regulation imposes a clear obligation upon designers and the CDM co-ordinator (in the case of notifiable projects) to take the general principles of prevention into account.

During the construction phase, regulation 7(2) applies which provides:

> Every person on whom a duty is placed by these Regulations in relation to the construction phase of the project shall ensure so far as is reasonably practicable that the general principles of prevention are applied in the carrying out of the construction work.

Regulation 7(2), at first glance, looks the same as regulation 7(1), except that it applies to the construction phase. However, the construction phase is qualified by the test of 'reasonably practicable'. It is not clear why the obligation on persons involved in the project prior to the construction phase are obligated to take account of the general principles of prevention, whereas those persons, who may be the same, involved in the construction phase are subject to the obligation of ensuring that the general principles of prevention are applied so far as is reasonably practicable. The Regulations are apparently inconsistent in this regard with respect to the general principles of prevention. However, it can probably be assumed that a very high threshold would be imposed by the court on those persons involved in the construction phase if they were to attempt to avoid the duty to apply the general principles of prevention by stating it was unreasonably practicable to do so.

Writing and documents

The definition of writing in regulation 2(1) includes:

> *writing which is kept in electronic form and which can be printed.*

The election of a client pursuant to regulation 8 requires that the person electing to be the client makes the agreement in writing. Writing is also referred to in regulation 14(5), which requires that the appointments made by the client for projects which are notifiable should be in writing. Although there are frequent references to notices, signs, plans and files, no other document is specifically required to be in writing, perhaps because such other documents could not be other than evidenced in writing.

All projects generate substantial numbers of documents that are part of the means for communication including the necessary co-operation and co-ordination. Above all else, documents are the evidence available to duty holders to prove that the duties under the Regulations have been discharged. Regulation 2(2) requires that the relevant documents are securely filed as follows:

> *Any reference in these Regulations to a plan, rules, document, report or copy includes a plan, rules, document or copy which is kept in a form –*
>
> *(a) in which it is capable of being reproduced as a printed copy when required; and*
>
> *(b) which is secure from loss or unauthorised interference.*

The importance of documents and a robust filing system is highlighted by the requirement to keep the health and safety file indefinitely. For larger and more complex projects, the harnessing of workflow and electronic document management systems would be a beneficial management tool.

5 Considerations for the welfare of workers

Introduction
Duties
Sanitary conveniences
Washing facilities
Drinking water
Changing rooms and lockers
Facilities for rest
First aid
Concluding remarks

Introduction

The 1994 Regulations did not include any provision for the health, safety and welfare of workers on construction sites, which had been dealt with in the Construction (Health, Safety and Welfare) Regulations 1996 and have now been revoked by the Regulations.

Schedule 2 of the Regulations sets out the requirements for the provision of welfare facilities to be provided by the contractors. The inclusion of welfare facilities within the Regulations is in line with the scope of the Temporary or Mobile Construction Sites Directive and finally recognises that the health and safety of workers is directly affected by their personal welfare.

The provision of a good standard of welfare facilities promotes recruitment, good morale and employee retention, which should be sufficient commercial justification for the investment in welfare facilities. While cold, overheated, dirty, dehydrated and uncomfortable workers are not going to be efficient and effective, they are also likely to be unsafe to themselves and their fellow workers.

The inclusion of the welfare facilities in the Regulations also recognises the wider issues that arise from public health and safety that would arise if there were inadequate sanitary facilities.

Duties

The duties upon the contractors in respect of welfare facilities for all projects are set out in regulation 13(7) as follows:

> *Every contractor shall ensure, so far as is reasonably practicable, that the requirements of Schedule 2 are complied with throughout the construction phase in respect of any person at work who is under his control.*

In respect of notifiable projects, the principal contractor also has a duty to ensure that Schedule 2 is observed as set out in regulation 22(1)(c):

> *ensure that welfare requirements sufficient to comply with the requirements of Schedule 2 are provided throughout the construction phase.*

Schedule 2 provides a specification for the welfare facilities, but does not go so far as to specify numbers of facilities or every detail. Indeed, this is a judgement left to the contractor because the Regulations recognise that it may not be practicably possible to provide all the welfare facilities.

However, the circumstances that would arise, whereby a contractor would have a defence that it was not reasonably practicable to provide welfare facilities required by Schedule 2, are likely to be very unusual. Of course, the contractor does not necessarily have to provide the welfare facilities him or herself, whereas in many cases the welfare facilities can be provided by arrangement, often in co-operation with the client or other contractors.

The client should be alert to ensure that the contractor or principal contractor provides the welfare facilities required by Schedule 2. The cost of welfare facilities can represent a disproportionate cost when compared to the cost of the construction works for small projects. The client should take due account of the fact that a contractor has allocated sufficient resources for welfare facilities or has taken steps to make arrangements with the client or adjacent land owners or others for use of other organisations' welfare facilities.

Evidence of the past record in planning and providing welfare facilities, and management of welfare facilities including the provision for cleaning and maintenance, is a core criterion for assessing competence by a person who is considering appointing a contractor.

Sanitary conveniences

Schedule 2 includes a requirement that construction sites are provided with appropriate sanitary conveniences as follows:

1. *Suitable and sufficient sanitary conveniences shall be provided or made available at readily accessible places. So far as is reasonably practicable, rooms containing sanitary conveniences shall be adequately ventilated and lit.*

2. *So far as is reasonably practicable, sanitary conveniences and the rooms containing them shall be kept in a clean and orderly condition.*

3. *Separate rooms containing sanitary conveniences shall be provided for men and women, except where and so far as each convenience is in a separate room the door of which is capable of being secured from the inside.*

The provision of sanitary conveniences prevents a public health risk and also preserves the modesty of both men and women.

The sanitary conveniences should be readily accessible and not involve a person having to walk or take other transport for a journey of an unreasonable duration. Construction sites should be planned so that the locations for sanitary conveniences are established at the outset.

The sanitary conveniences should be adequately ventilated and lit, which is particularly relevant if the construction work is taking place during hours of darkness.

Sanitary conveniences that are not kept in a clean and orderly condition are likely to present a health hazard.

The workforce on a construction site is now almost certainly no longer entirely made up of men, and therefore men and women should be catered for separately unless that is not reasonably practicable, in which case the doors should be capable of being locked from the inside.

Washing facilities

The washing facilities that are required to be provided by a contractor are set out in Schedule 2 as follows:

4. *Suitable and sufficient washing facilities including showers if required by the nature of the work or for health reasons, shall so far as is reasonably practicable be provided or made available at readily accessible places.*

5. *Washing facilities shall be provided –*

    (a) in the immediate vicinity of every sanitary convenience, whether or not provided elsewhere; and

(b) in the vicinity of any changing rooms required by paragraph 15 whether or not provided elsewhere.

6. *Washing facilities shall include –*

(c) a supply of clean hot and cold, or warm, water (which shall be running water so far as is reasonably practicable);

(d) soap or other suitable means of cleaning; and

(e) towels or other suitable means of drying.

7. *Rooms containing washing facilities shall be sufficiently ventilated and lit.*

8. *Washing facilities and the rooms containing them shall be kept in a clean and orderly condition.*

9. *Subject to paragraph 10 below, separate washing facilities shall be provided for men and women, except where and so far as they are provided in a room the door of which is capable of being secured from inside and the facilities in each such room are intended to be used by only one person at a time.*

10. *Paragraph 9 above shall not apply to facilities which are provided for washing hands, forearms and face only.*

In planning for washing facilities the contractor should take account of the nature of the construction work being undertaken by the workers, particularly activities that cover workers in mud and dust. The Regulations recognise that washing facilities may only be necessary for washing hands, forearms and face only. In those circumstances it is not necessary to provide separate washing facilities for men and women. Where, however, washing facilities are necessary for washing other parts of the body or showering, separate facilities for men and women should be provided.

When planning the layout for sanitary conveniences, due account should be taken of the provision of washing facilities.

Drinking water

The requirements for drinking water set out in Schedule 2 are as follows:

11. *An adequate supply of wholesome drinking water shall be provided or made available at readily accessible and suitable places.*

12. *Every supply of drinking water shall be conspicuously marked by an appropriate sign where necessary for reasons of health and safety.*

> *13. Where a supply of drinking water is provided, there shall also be provided a sufficient number of suitable cups or other drinking vessels unless the supply of drinking water is in a jet from which persons can drink easily.*

The water must be safe to drink and made available at readily accessible and suitable places on the construction site. On some construction sites, where water is recycled for use in the sanitary conveniences and washing, it is important that water which is safe to drink is clearly identified.

The contractor is required to provide drinking vessels for the works unless the water is under pressure and can be drunk from a jet.

The need to provide water for drinking facilities may be a particular challenge for remote construction sites. Drinking water can be supplied in bowsers, in which case the contractor should ensure that there is an adequate supply based on the proper estimates of quantities that will be consumed. The Food Standards Agency recommends at least 1.2 litres of fluids each day for normal conditions. Fluid requirements will increase as the ambient temperature increases and as the physical effort expended by workers increases.

Changing rooms and lockers

Construction workers need facilities to change into appropriate clothing for the work to be undertaken and Schedule 2 requires changing rooms and lockers to be provided as follows:

> *14(1) Suitable and sufficient changing rooms shall be provided or made available at readily accessible places if –*
>
> > *(a) a worker has to wear special clothing for the purposes of his work; and*
> >
> > *(b) he cannot, for reasons of health or propriety, be expected to change elsewhere;*
>
> *being separate rooms for, or separate use of rooms by, men and women where necessary for reasons of propriety.*
>
> *(2) Changing rooms shall –*
>
> > *(a) be provided with seating;*
> >
> > *(b) include, where necessary, facilities to enable a person to dry any such special clothing and his own clothing and personal effects.*

(3) Suitable and sufficient facilities shall, where necessary, be provided or made available at readily accessible places to enable persons to lock away –

(a) any such special clothing which is not taken home;

(b) their own clothing which is not worn during working hours; and

(c) their personal effects.

The requirement to provide changing rooms and lockers may be a significant expense for small projects. However, the need for workers to have somewhere secure to store their clothing and other personal belongings and dry work clothes should not be overlooked or underestimated.

Facilities for rest

Workers are entitled to breaks during the working day for taking refreshment or for receiving medical assistance. The facilities for rest are set out in Schedule 2 as follows:

15(1) Suitable and sufficient rest rooms or rest areas shall be provided or made available at readily accessible places.

(2) Rest rooms and rest areas shall –

(a) include suitable arrangements to protect non-smokers from discomfort caused by tobacco smoke;

(b) be equipped with an adequate number of tables and adequate seating with backs for the number of persons at work likely to use them at any one time;

(c) where necessary, include suitable facilities for any person at work who is a pregnant woman or nursing mother to rest lying down;

(d) include suitable arrangements to ensure that meals can be prepared and eaten;

(e) include the means for boiling water; and

(f) be maintained at an appropriate temperature.

The Regulations state specifically that the rest facilities should provide the opportunity for workers to eat meals and have access to boiling water. The rest facilities are also required to be maintained at an appropriate temperature which may mean air-conditioning or fan-assisted air circulation in extreme hot conditions in addition to the normal requirement for heating.

First aid

First aid facilities are not specifically referred to in the Regulations, although the ACOP refers to first aid together with the welfare facilities in the contents for the construction phase plan. The welfare facilities are, at the very least, required to have suitable facilities for a pregnant woman or a nursing mother in which case the extension to first aid facilities generally is logical.

Concluding remarks

The welfare facilities are an essential part of planning and managing construction work. The principal contractor, in the case of notifiable projects and other contractors, should take account of the installation of welfare facilities as part of the overall programme. Welfare facilities should be available at the outset of construction work and the client should take steps to ensure that welfare facilities are provided to an acceptable level capable of delivering appropriate facilities for the nature of the construction work. Contractors should also make provision for the cleaning and maintenance of welfare facilities throughout the entire duration of the construction work.

6   Consideration for hazardous operations

Introduction

Part 4 has direct relevance to working on a construction site and the identification of specific hazards as opposed to the administration and management of health and safety.

An important and overlooked benefit of Part 4 is the very fact that it provides a checklist against which a designer and a contractor can assess the inherent risks in the project and its design and construction relative to each of the regulations 26 to 44 inclusive.

Duties relating to health and safety on construction sites

The duties on clients, contractors, designers, CDM co-ordinators, including organisations and individuals to ensure that all the hazards and activities in regulations 26 to 44 are taken account of, are described in regulation 25 which provides:

> *(1) Every contractor carrying out construction work shall comply with the requirements of regulations 26 to 44 insofar as they affect him or any person carrying out construction work under his control or relate to matters within his control.*

> *(2) Every person (other than a contractor carrying out construction work) who controls the way in which any construction work is carried out by a person at work shall comply with the requirements of regulations 26 to 44 insofar as they relate to matters which are within his control.*

It follows that it is imperative that every contractor and other person who has any control over the manner in which persons undertake any tasks on a construction site must be familiar with regulations 26 to 44. The subject matter of regulations 26 to 44 provides the ideal checklist for identification of risks and the basis for a training programme for all persons immediately prior to working on a site.

Any person on a construction site has a duty to themselves and every other person on site, underlined by regulation 25(3) of the Regulations, which provides:

> *Every person at work on construction work under the control of another person shall report to that person any defect which he is aware may endanger the health and safety of himself or another person.*

A person is under a further positive duty to report any defect of which he or she may be aware could cause a danger. A person cannot be expected to recognise certain types or specific defects as likely to be dangerous without appropriate training. This must be part of an education and training programme for site personnel on a continuous basis.

Regulation 25(1) and (2) do not apply to regulation 33 when other persons are identified by regulation 25(4) as follows:

> *Paragraphs (1) and (2) shall not apply to regulation 33, which expressly says on whom the duties in that regulation are imposed.*

Safe places of work

The fundamental principle in relation to safe places of work as found in regulation 26(2) is put as simply as:

> *Every place of work shall, so far as is reasonably practicable, be made and kept safe for, and without risks to health to, any person at work there.*

Place of work is defined in regulation 2(1) as meaning:

> *any place which is used by any person at work for the purposes of construction work or for the purposes of any activity arising out of or in connection with construction work.*

The definition is more comprehensive than might be understood by many to be a construction site which is defined as including:

> *any place where construction work is being carried out or to which the workers have access, but does not include a workplace within it which is set aside for purposes other than construction work.*

While the place of work will mean a construction site, it also denotes a specific place where a person works, e.g. within a cofferdam, or off site where prefabrication or pre-casting yards are supplying the construction site. Crucially, the place of work, as referred to in regulation 26(2), will change in material respects with time as the construction work proceeds. It is vital that the changing situation at a place of work is recognised as a hazard in itself and therefore requires that places of work need to be monitored for changing risks and hazards. To summarise the effect of regulation 26 it deals with:

1. safe access and egress; and

2. keeping the workplace safe including the provision of adequate working space.

In complying with regulation 26 it is virtually impossible not to have regard to regulations 27 to 32 and 34 to 44, which all contribute to providing a safe place of work. There are broad categories of compliance which can be sub-divided as in Table 6.1.

Table 6.1   Categories of compliance

Site movement	Traffic routes (36), vehicles (37), emergency procedures (39), emergency routes and exits (40), lighting (44)
Construction of structures	Stability of structures (28), excavations (31), cofferdams and caissons (32)
Site environment	Good order and site security (27), fresh air (42), temperature and weather protection (43), lighting (44)
Specific hazards	Good order and site security (27), demolition or dismantling (29), explosives (30), energy distribution installations (34), prevention and drowning (35), prevention of risk from fire, etc. (38)

The regulations collected together under the above categories have only been grouped for convenience and as an aid to the planning of the construction phase. There is no intention to exclude any other regulations from consideration under a category of compliance but it makes more sense of what first appears to be a random selection and order of topics.

Site movement

The need to move to and from one place to another on a construction site without risk is a fundamental requirement. Regulation 26(1) states:

> *There shall, so far as is reasonably practicable, be suitable and sufficient safe access to and egress from every place of work and to and from every other place provided for the use of any person while at work, which access and egress shall be properly maintained.*

Always subject to the principle of so far as reasonably practicable, the reference to 'suitable and sufficient' means that the person(s) in regulation 25 have to take account of size, weight and frequency of usage that personnel, vehicles, equipment and materials will need for access and egress. The need to maintain safe access and egress is an express obligation, as is the need to prevent any person using unplanned access or egress as underlined by regulation 26(3) which states:

Suitable and sufficient steps shall be taken to ensure, so far as is reasonably practicable, that no person uses access or egress, or gains access to any place, which does not comply with the requirements of paragraph (1) or (2) respectively.

Traffic routes

The Regulations recognise that access and egress and movement within a construction site require planning and that pedestrians must co-exist with vehicular traffic. A traffic route is defined as meaning:

a route for pedestrian traffic or for vehicles and includes any doorway, gateway, loading bay or ramp.

A loading bay is defined as meaning:

any facility for loading or unloading.

Regulation 36 sets out in detail a list of 'do's' to ensure compliance with the requirements for traffic routes. The basic obligations are in regulations 36(1) and (2) which state:

(1) Every construction site shall be organised in such a way that, so far as is reasonably practicable, pedestrians and vehicles can move safely and without risks to health.

(2) Traffic routes shall be suitable for the persons or vehicles using them, sufficient in number, in suitable positions and of sufficient size.

Regulation 36(3) states that sub-paragraph (2) cannot be satisfied unless points (a) to (e) are observed and/or provided as follows:

A traffic route shall not satisfy sub-paragraph (2) unless suitable and sufficient steps are taken to ensure that –

(a) pedestrians or vehicles may use it without causing danger to the health or safety of persons near it.

To fulfil this requirement there will need to be, at the very least, a risk assessment of the likelihood of persons being too near a traffic route such that they would be in harm's way.

(b) any door or gate for pedestrians which leads onto a traffic route is sufficiently separated from that traffic route to enable pedestrians to see any approaching vehicle or plant from a place of safety.

This requires there to be a sufficient visibility splay for pedestrians to assess the distance and speed of any approaching vehicle before crossing the route.

(c) there is sufficient separation between vehicles and pedestrians to ensure safety or, where this is not reasonably practicable –

(i) there are provided other means for the protection of pedestrians, and

(ii) there are effective arrangements for warning any person liable to be crushed or trapped by any vehicle of its approach.

Sufficient separation will depend upon size and speed of vehicles and quality of surface to avoid vehicles straying onto the pedestrian routes. A risk assessment should consider splashes, spillages and jettisoning of loads that could endanger pedestrians. Effective arrangements for warning persons as required by (c)(ii) should include audible warnings and sensors that engage when the vehicle is in reverse gear.

Sub-paragraphs (d) and (e) are self-explanatory and prescriptive as follows:

(d) any loading bay has at least one exit point for the exclusive use of pedestrians; and

(e) where it is unsafe for pedestrians to use a gate intended primarily for vehicles, one or more doors for pedestrians is provided in the immediate vicinity of the gate, is clearly marked and is kept free from obstruction.

Traffic routes have to be obvious and maintained such that they do not become health and safety hazards throughout the terms of the construction work as required by regulation 26(4) which provides that:

(4) Every traffic route shall be –

(a) indicated by suitable signs where necessary for reasons of health or safety;

(b) regularly checked; and

(c) properly maintained.

Maintenance of traffic routes should involve maintaining the trafficked surface but it also follows that traffic routes must be unobstructed and capable of being trafficked as provided for in regulation 36(5) which states:

(5) No vehicle shall be driven on a traffic route unless, so far as is reasonably practicable, that traffic route is free from obstruction and permits sufficient clearance.

Vehicles

Regulation 37(1) states:

> *(1) Suitable and sufficient steps shall be taken to prevent or control the unintended movement of any vehicle.*

A vehicle is defined as including:

> *any mobile work equipment*

and work equipment is defined as meaning:

> *any machinery, appliance, apparatus, tool or installation for use at work (whether exclusively or not).*

Vehicles will include, therefore, trucks, cars, mobile cranes, excavators, drilling rigs, concrete pumps, mobile generators, fuel tenders, in fact any item on a construction site that can be moved under its own power or towed.

Unintended movement of a vehicle needs to be considered both during intended movement and when stationary. Unintended movement when actually on the move can be caused by (a) ground instability, (b) excessive gradients, (c) deflection or failure of a supporting structure, (d) undue speed on an irregular driving surface, (e) punctures, and (f) competence of the driver/operator.

The risk of an unintended movement when stationary can be caused by (a) to (c) but can also be caused by vandalism or theft. The prevention or mitigation of unintended movement can be addressed by considering each of the causes and introducing appropriate measures. Drivers should have to be competent by virtue of regulation 4(1)(c)(i) and such persons not having passed an appropriate test to be in control of a vehicle would be a breach of regulation 37(1) and an immediate hazard.

There are a large number of accidents on construction sites caused by persons being knocked down or crushed by vehicles and regulation 37(2) requires that a vehicle has a means of warning. Regulation 37(2) states:

> *(2) Suitable and sufficient steps shall be taken to ensure that, where any person may be endangered by the movement of any vehicle, the person having effective control of the vehicle shall give warning to any person who is liable to be at risk from the movement of the vehicle.*

The person who has effective control of the vehicle has the onus of warning pedestrians. In each case the responsible person in regulation 25 will have to decide what means would be appropriate for warning pedestrians, e.g. audible and/or flashing light warnings and whether entirely manual or automatic. The duty under regulation 37(2) should be related to the measures taken to comply with regulation 36(3)(c)(ii).

Regulation 37(3) goes to the issue of competence. Only a properly trained driver can be expected to ensure compliance with the requirements of regulation 37(3) as follows:

> *(3) Any vehicle being used for the purposes of construction work shall when being driven, operated or towed –*
>
> > *(a) be driven, operated or towed in such a manner as is safe in the circumstances; and*
> >
> > *(b) be loaded in such a way that it can be driven, operated or towed safely.*

Regulation 37(4) specifically bans unintended passengers joyriding or hanging on for a lift, which provides as follows:

> *(4) No person shall ride or be required or permitted to ride on any vehicle being used for the purposes of construction work otherwise than in a safe place thereon provided for that purpose.*

Persons who ride without permission may be risking their own safety and the safety of others and as such should be subject to disciplinary measures.

Loading or unloading of loose materials, which include aggregates and powders, have specific hazards and regulation 37(5) emphasises the need for a person to be in a safe place of work during such loading or unloading as follows:

> *(5) No person shall remain or be required or permitted to remain on any vehicle during the loading or unloading of any loose material unless a safe place of work is provided and maintained for such person.*

The risks referred to in 37(6) might more appropriately have been listed under regulation 36.

> *(6) Suitable and sufficient measures shall be taken so as to prevent any vehicle from falling into any excavation or pit, or into water, or overrunning the edge of any embankment or earthwork.*

The sufficient and suitable measures will include width of the route, adequate slope and/or excavation, stability, kerbs and/or barriers.

The matters listed in regulation 39(2) are exactly the same matters which, it is suggested, should be used to assess safe access to and egress from a construction site.

These are set out as follows:

(a) *the type of work for which the construction site is being used;*

(b) *the characteristics and size of the construction site and the number and location of the places of work on that site;*

(c) *the work equipment being used;*

(d) *the number of persons likely to be present on the site at one time; and*

(e) *the physical and chemical properties of any substances or materials on or likely to be on that site.*

Emergency procedures

Regulation 39 is concerned with emergency procedures, while Regulation 40 is concerned with emergency routes and exits. The relevant fire authority may be the enforcing authority in respect of this aspect of the works, and consulting with the relevant fire authority may be appropriate for large and complex construction sites.

Regulation 39(1) creates the duty to make arrangements for emergency procedures as follows:

Where necessary in the interests of the health and safety of any person on a construction site, there shall be prepared and, where necessary, implemented suitable and sufficient arrangements for dealing with any foreseeable emergency which arrangements shall include procedures for any necessary evacuation of the site or any part thereof.

In making the arrangements, account should be taken of the points under regulation 39(2)(a) to (e), as above. Foreseeable emergencies shall vary according to the location of the construction site and proximity to man-made or natural hazards.

Arrangements made for emergency procedures must be communicated to the persons who will rely upon the emergency procedures being effective. The procedures should also be tested to ensure that they are effective. Regulation 39(3) creates such duties as follows:

Where arrangements are prepared pursuant to paragraph (1), suitable and sufficient steps shall be taken to ensure that –

(a) *every person to whom the arrangements extend is familiar with those arrangements; and*

(b) *the arrangements are tested by being put into effect at suitable intervals.*

Emergency routes and exits

The emergency routes and exits are an integral part of the emergency procedures. An emergency route is designed to evacuate persons at risk away from the source of danger as quickly as possible. Regulation 40(1) makes the point commensurate with the duty to provide such emergency routes as follows:

> *Where necessary in the interests of the health and safety of any person on a construction site, a sufficient number of suitable emergency routes and exists shall be provided to enable any person to reach a place of safety quickly in the event of danger.*

The emergency routes do not have to follow the established traffic routes. However, the remainder of regulation 40 in paragraphs (2) to (5) provides rules that are an integral component of an emergency route and exit as follows:

> *(2) Any emergency route or exit provided pursuant to paragraph (1) shall lead as directly as possible to an identified safe area.*

> *(3) Any emergency route or exit provided in accordance with paragraph (1), and any traffic route giving access thereto, shall be kept clear and free from obstruction and, where necessary, provided with emergency lighting so that such emergency route or exit may be used at any time.*

> *(4) In making provision under paragraph (1) account shall be taken of the matters in regulation 39(2).*

> *(5) All emergency routes or exits shall be indicated by suitable signs.*

Therefore, emergency routes and exits must:

(i)   be as short as possible leading to safe areas;

(ii) be kept clear of obstructions;

(iii) be provided with emergency lighting;

(iv) take account of the matters that were assessed for the emergency procedures;

(v)   be identified by suitable signs.

Lighting

Traffic and emergency routes and movement on site can only be safe and compliant with the Regulations if visibility is clear in all situations.

In hours of darkness, the need for lighting is obvious, but lighting will also be necessary when working in unlit interiors, underground, and heavy shadows during daylight hours. For this reason, regulation 44 requires the following:

> *(1) Every place of work and approach thereto and every traffic route shall be provided with suitable and sufficient lighting, which shall be, so far as is reasonably practicable, by natural light.*

> *(2) The colour of any artificial lighting provided shall not adversely affect or change the perception of any sign or signal provided for the purposes of health and safety.*

> *(3) Without prejudice to paragraph (1), suitable and sufficient secondary lighting shall be provided in any place where there would be a risk to the health and safety of any person in the event of failure of any primary artificial lighting.*

To ensure compliance with regulation 44, the light source should not be of a wavelength which changes colours or the ability of any person to read the signs or signals that have been provided for health and safety purposes.

The lighting will usually be powered from the mains electricity supply. However, the risk of power cuts is foreseeable and where artificial lighting is required, especially for underground works and unlit interiors, there should be a provision for emergency generation. This requirement is a provision that should be highlighted in the construction phase plan.

Site environment

Good order and site security

The site environment includes space, tidiness, security, fresh air, general climate and visibility. All the factors that contribute to a safe environment can be planned for and provided by the design and construction planning.

Working space is important so that the workers' tasks can be carried out safely and without compromise, and quick avoidance of immediate danger. Regulation 26(4) requires that:

> *Every place of work shall, so far as is reasonably practicable, have sufficient working space and be so arranged that it is suitable for any person who is working or is likely to work there, taking account of any necessary work equipment present.*

Regulation 26(4) draws attention to the fact that the work equipment will also be taking up space. This may include hand tools in addition to freestanding equipment. Importantly, the equipment should not prevent free movement or access to egress from the place of work.

An untidy site will invariably be a more dangerous site. One of the common injuries to workers is injuries to the foot caused by upstanding nails or other sharp objects. Regulation 27(1) requires that:

> *Every part of the construction site shall, so far as is reasonably practicable, be kept in good order and every part of the construction site which is used as a place of work shall be kept in a reasonable state of cleanliness.*

There can be no excuse for a contractor to allow a construction site to be a risk to health and safety as a result of plant, materials and debris not being kept in their proper place and/or being removed. The tidiness that exists on a construction site is a good indicator of the attitude of a contractor to health and safety and the culture that is likely to exist towards the management of health and safety generally.

The importance of preventing injuries to feet is highlighted by regulation 27(3) as follows:

> *No timber or other material with projecting nails (or similar sharp objects) shall –*
>
> *(a) be used in any work; or*
>
> *(b) be allowed to remain in any place,*
>
> *if the nails (or similar sharp object) may be a source of danger to any person.*

Contractors have a duty pursuant to regulation 13(6) to prevent unauthorised persons entering upon their construction site. Regulation 27(2) places a specific duty upon a contractor to secure the perimeter of the construction site as follows:

> *Where necessary in the interests of health and safety, a construction site, so far as is reasonably practicable and in accordance with the level of risk posed, either*
>
> *(a) have its perimeter identified by suitable signs and be so arranged that its extent is readily identifiable; or*
>
> *(a) be fenced off,*
>
> *or both.*

The contractor has, therefore, to undertake a risk assessment as to the required degree of security at the perimeter. On sites where the hazards represent a severe risk to health and safety, the precautions that a contractor should take to prevent unauthorised entry might include the need to invest in high steel fencing topped with barbed wire or other means of preventing climbing over together with coverage of the perimeter by CCTV. At another extreme, on a site of low risk the security to the site might only require some signs and temporary taping.

Fresh air

Fresh air includes air that can be breathed without harm to health. Thus, fresh air will not have noxious or dangerous fumes, or a high dust load of fine particulate matter. Regulation 42 requires the following steps:

> *Suitable and sufficient steps shall be taken to ensure, so far as is reasonably practicable, that every place of work or approach thereto has sufficient fresh or purified air to ensure that the place or approach is safe and without risks to health.*

The steps that can be taken to ensure that the fresh air is not polluted by noxious or dangerous fumes or airborne particulate matter include fans, air-conditioning and dust suppression. If such systems are provided then it is important that account is taken of the consequences of any failure of such system as provided for by regulation 42(2):

> *Any plant used for the purpose of complying with paragraph (1) shall, where necessary, for reasons of health and safety, include an effective device to give visible or audible warning of any failure of the plant.*

In the event of such failure, the workers should have access to personal protective equipment, including goggles, face masks and personal breathing apparatus, especially if working underground.

Temperature and weather protection

For workers engaged in construction activities indoors, they should not be exposed to excessive heat or cold. Regulation 43(1) requires that:

> *Suitable and sufficient steps shall be taken to ensure, so far as is reasonably practicable, that during working hours the temperature at any place of work indoors is reasonable having regard to the purpose for which that place is used.*

In undertaking a risk assessment of the temperature range likely to be experienced indoors and the wellbeing of workers, account should be taken of the nature of the construction activities. Where heavy physical work is needed, the maximum acceptable temperature will be lower than what might be needed for fine manual movement. Conversely, the lowest acceptable temperature may be lower. The contractor should consider heating and air-conditioning or other means of circulating air. In the event that this is not reasonably practicable, the contractor should issue appropriate personal protective equipment to the workers and also consider the maximum amount of time that any worker should be exposed to such conditions in a shift.

The external environment cannot be controlled. However, by providing shelter, which can be created by the planned sequence of construction of the permanent works, workers can be afforded some protection. To that extent, regulation 43(2) provides that:

> *Every place of work outdoors shall, where necessary to ensure the health and safety of persons of work there, be so arranged that, so far as is reasonably practicable and having regard to the purpose for which that place is used and any protective clothing or work equipment provided for the use of any person who works there, it provides protection from adverse weather.*

Although the contractor can have taken such steps as are reasonably practicable in providing temporary shelter or shelter from the permanent works and having issued appropriate personal protective clothing, the adverse weather conditions might still be considered a risk to health and safety. In such circumstances, the contractor has no option other than to suspend the works until such time as the adverse weather conditions abate.

Lighting

The conditions on a construction site should be clearly visible and this will require artificial lighting outside of daylight hours. Lighting has been discussed above, under the category of site movement.

Construction of structures

Stability of structures

Structure is a defined term and comes within the scope of regulation 28.

If structures are unstable they obviously present a risk to the health and safety of workers and other persons who may be affected by the construction work. Regulation 28 which makes provision for the stability of structures relates to the construction phase and does not address stability post construction.

The construction of structures should not present a risk to the health and safety of workers on the construction site. Equally, existing structures that are required to be modified should also not present any risks arising from instability. Regulation 28 creates a duty to ensure that structures are stable as follows:

> *(1) All practicable steps shall be taken, where necessary to prevent danger to any person, to ensure that any new or existing structure or any part of such structure which may become unstable or in a temporary state of repair or instability due to the carrying out of construction work does not collapse.*

> *(2) Any buttress, temporary support or temporary structure must be of such design or so installed or maintained as to withstand any foreseeable loads which may be imposed on it and must only be used for the purpose for which it is so designed, installed and maintained.*

> *(3) No part of a structure shall be so loaded as to render it unsafe to any person.*

The reference to structures includes existing permanent structures, new structures under construction and structures associated with temporary works.

The role of the designer in identifying potential states of instability cannot be overstated. The starting-point for any risk assessment associated with transient instability or weakness is the identification of the foreseeable loads. Both the designer and the contractor should liaise to identify loads that can be created by the construction process due to plant and equipment, spoil heaps, the filling of tanks, exposure to wind loading and so on.

After establishing the necessary loading conditions the designer should consider the condition of the structure. Existing structures may be weakened by partial demolition, materials may not have acquired full strength, particularly reinforced concrete, and assumptions should be checked that were made at the design stage about dimensions that could not be verified before construction.

In respect of temporary structures, the parties should have exchanged design information to ensure that the assumptions for loading are communicated and that appropriate steps taken not to exceed the loading during construction.

Excavations

Regulation 2(1) defines 'excavation' as including:

> *any earthwork, trench, well, shaft, tunnel or underground working.*

One of the areas in which clients can make a significant contribution in providing resources for improving health and safety management of excavation operations is allocating sufficient funds for site investigation. Too often the site investigation is underfunded, if funded at all, by the client. It is incumbent upon the designers and the contractors to satisfy themselves that the information they require for design includes the ground information, so that excavations can be undertaken safely. Where there are significant excavations required as part of the project, a client should be prepared to allocate further resources for site investigation during the course of construction if the predicted soil or rock conditions, as exposed by the excavation, differ from the conditions predicted from the original site investigation.

Regulation 31 refers to the duty to prevent danger to any person from excavation by requiring all practicable steps to have been taken. These practicable steps include the site investigation.

Regulation 31(1) states as follows:

> *(1) All practicable steps shall be taken, where necessary to prevent danger to any person, including, where necessary, provision of supports or battering, to ensure that –*
>
> *(a) any excavation or part of any excavation does not collapse;*
>
> *(b) no material from a side or roof of, or adjacent to, any excavation is dislodged or falls; and*
>
> *(c) no person is buried or trapped in an excavation by material which is dislodged or falls.*

The design of supports or battering to an excavation should never be guess work. The provision of supports or battering is a design process that requires knowledge of the relevant soil or rock parameters. These can only be obtained from appropriate in situ and laboratory testing as part of the site investigation.

The appropriate design of linings to tunnels, wells, shafts and the sides of open earthwork excavations should prevent the dislodgement or falling of material.

Open excavations are an obvious hazard to workers on a construction site. Such hazards are a particular risk to persons who obtain un-authorised entry to construction sites. Regulation 31(2) requires that:

> *Suitable and sufficient steps shall be taken to prevent any person, work equipment, or any accumulation of material from falling into any excavation.*

The steps to be taken are to prevent three occurrences. First, any person falling into an excavation, second, work equipment falling into an excavation and, finally, any spoil from the excavation falling back into the excavation. These steps will include physical barriers of appropriate strength and dimensions, and will also include restrictions on traffic and site movement adjacent to the excavations, and site planning so that stockpiling of materials or excavated spoil are sufficiently distant from the edge of the excavation.

The access of plant and equipment close to excavations cannot always be avoided, in which case the designer is required to ensure that the support to the excavation can withstand such loading. Regulation 31(3) requires:

> *Without prejudice to paragraphs (1) and (2), suitable and sufficient steps shall be taken, where necessary, to prevent any part of an excavation or ground adjacent to it from being overloaded by work equipment or material.*

If work equipment or material is needed close to the excavation, the designer will need confidence in the soil or rock properties to factor in a sufficient factor of safety in the event of inadvertent loads being placed adjacent to the excavation.

Excavations are a serious hazard for workers on construction sites. Many construction workers have been killed working in trenches that are not as deep as the height of the victim. A construction worker who goes to work in an excavation to carry out construction activity has the right to be confident that it is a safe place to work. Regulation 31(4) recognises the serious hazards associated with excavations by requiring that excavations should be inspected by a competent person as follows:

> *Construction work shall not be carried out in an excavation where any supports or battering have been provided pursuant to paragraph (1) unless –*
>
> *(a) the excavation and any work equipment and materials which affect its safety, have been inspected by a competent person –*

    (i)   *at the start of the shift in which the work is to be carried out,*

    (ii) *after any event likely to have affected the strength or stability of the excavation, and*

    (iii) *after any material unintentionally falls or is dislodged; and*

  (b) *the person who carried out the inspection is satisfied that the work can be carried out there safely.*

The requirement for inspection by a competent person only arises where the excavation is either supported or where the profile of the edge of excavation has been cut to a slope. The condition of the support or the angle and profile of the slope within the excavation has to be inspected. The requirements with respect to the reports of inspections are dealt with in regulation 33. However, the significance of the requirement that the inspection should be carried out by a competent person with respect to excavations is particularly difficult to assess. A person can inspect the supports or slopes and identify any movement or distortion to the profile at the beginning of the shift. Such an inspection may, of course, include monitoring of measurements to reference points, but will the same person know what to do if there is any change in the measurements? In the opinion of the author, competence cannot be achieved without the inspector having a sound grasp of geotechnical engineering for anything other than limited excavations for isolated foundations. Events that are likely to affect the strength or stability of the excavation are unlikely to be known to persons who are not educated to a high level in geotechnical engineering. To identify such events requires close inspection and monitoring throughout the excavation, to the extent that a competent geotechnical engineer may have assessed a need for an in situ monitoring system to check for pore pressures and soil movements.

Only after such checks have taken place before the start of a shift and the person referred to in regulation 31(4)(b) is satisfied that the work can proceed safely, should work proceed.

In the event that the person who has undertaken the inspection required by regulation 31(4) has any concerns, regulation 31(5) provides that:

> *Where the person who carried out the inspection has under regulation 33(1)(a) informed the person on whose behalf the inspection has been carried out on any matter about which he is not satisfied, work shall not be carried out in the excavation until the matters have been satisfactorily remedied.*

The person who has undertaken the inspection is likely to be aware of the commercial consequences of preventing further excavation work. For that reason, it is especially important that the competent person is properly trained in geotechnical principles, and has sufficient seniority and experience not to be unduly influenced by the commercial consequences when health and safety matters associated with excavations are at stake.

Cofferdams and caissons

Cofferdams and caissons are usually associated with heavy civil engineering works. The fundamental requirement of a cofferdam or caisson is to provide a working environment within a structure that is retaining soil or water on the outside. The design of a cofferdam or caisson requires the same extent of knowledge as required by excavations if the cofferdam or caisson is retaining soil; whereas, in the case of a cofferdam or caisson that will be retaining water to maintain a dry working environment, the designer should take account of conditions that will vary the depth of water. The consequences of a failure in a cofferdam or caisson in which workers are engaged in construction work are sufficiently serious to devote a considerable amount of time and expense in complying with regulation 32(1) which requires that:

Every cofferdam or caisson shall be –

(a) of suitable design and construction,

(b) appropriately equipped so that workers can gain shelter or escape if water or materials enter it, and

(c) properly maintained.

Maintaining the cofferdam or caisson may involve supplementary construction work such as grouting, putting in place additional structural members, caulking or the installation of pumping to remove water.

The means of equipping the cofferdam or caisson so that workers can gain shelter or escape if the retained water or material on the outside enters the working environment will depend on a number of factors – including the depth of the excavation within the cofferdam or caisson – the mode of failure – including water bursts or overtopping – and whether soil or water is being retained.

The risk to health and safety of any persons working in a cofferdam or caisson is similar to the risk of working in excavations, such that regulations require that a competent person inspects the cofferdam or caisson in accordance with regulation 32(2) which requires –

A cofferdam or caisson shall be used to carry out construction work only if –

(a) the cofferdam or caisson, and any work equipment and materials which affect its safety, have been inspected by a competent person,

    (i) at the start of the shift in which the work is to be carried out, and

    (ii) after any event likely to affect the strength or stability of the cofferdam or caisson; and

(b) the person who carried out the inspection is satisfied that the work can be safely carried out there.

The extent and scope of the inspection is very similar to the requirements for inspection for excavations. If a cofferdam or caisson is within soil the competent person should be similarly qualified as the competent person to inspect excavations. In respect of cofferdams or caissons that are in water, the competent person will be a civil engineer with experience of the design and construction of cofferdams and caissons in such circumstances. Regulation 32(3) is in the same terms as regulation 31(5) for excavations, such that if the competent person is not satisfied that the cofferdam or caisson can provide a safe place of work then work should not be carried out within the cofferdam or caisson.

Reports of inspections

The importance of ensuring the integrity of excavations, cofferdams and caissons, pursuant to regulations 31 and 32 respectively, is such that the regulations have required a competent person to make inspections at the start of every shift. Regulation 33 provides the regulatory requirements for the inspection procedure.

Reports will be the evidence that the regulations have been followed and moreover they will also be the evidence needed to investigate the cause of any failures in excavations, cofferdams or caissons. Although the competent person is required to inspect the excavation, cofferdams and caissons at the start of every shift, only one report needs to be completed every 7 days, as required by regulation 33(5) as follows:

Nothing in this regulation shall require as regards an inspection carried out on a place of work for the purposes of regulations 31(4)(a)(i) and 32(2)(a)(i), the preparation of more than one report within a period of 7 days.

Regulation 33(1) is primarily addressed to the person who carries out the inspection and provides as follows:

> *Subject to paragraph (5), the person who carries out an inspection under regulation 31 or 32 shall, before the end of the shift within which the inspection is completed –*
>
> *(a) where he is not satisfied that the construction work can be carried out safely at the place inspected, inform the person for whom the inspection was carried out of any matters about which he is not satisfied; and*
>
> *(b) prepare a report which shall include the particulars set out in schedule 3.*

In the event that the inspector is not satisfied that the construction work can be carried out safely, he or she should draw this to the attention of the person for whom the inspection is carried out without delay and before the end of the shift within which the inspection was completed. In any event, the inspector is required to prepare a report to include the particulars in Schedule 3 which are:

1. Name and address of the person on whose behalf the inspection was carried out.

2. Location of the place of work inspected.

3. Description of the place of work or part of that place inspected (including any work equipment and materials).

4. Date and time of inspection.

5. Details of any matter identified that could give rise to a risk to the health or safety of any person.

6. Details of any action taken as a result of any matter identified in paragraph 5 above.

7. Details of any further action considered necessary.

8. Name and position of the person making the report.

The report has to be provided within 24 hours of the inspection and given to the person who was required to obtain the inspection as required by regulation 33(2) as follows:

> *A person who prepares a report under paragraph 1 shall, within 24 hours of completing the inspection to which the report relates, provide the report or a copy of it to the person on whose behalf the inspection was carried out.*

The person who is required to prepare the report has only 24 hours in which to provide it for projects which involve significant excavations, cofferdam or caissons there will be many days, weeks or even months of reporting. The requirement for a competent person to undertake the inspection should take account of cover for absence by an inspector.

The last report should be prepared when either the excavation is completed to its final profile without any permanent works or at such time when the permanent works are completed within the excavation. In the case of cofferdams and caissons the last day of inspection will be on the day that the cofferdam or caisson is dismantled or incorporated into the permanent works.

The person undertaking the inspection can be an external consultant. By this means the person for whom the inspection is being carried out can ensure that the appropriate level of competence is obtained. The employer or person under whose control the inspector carries out the work is under a duty to ensure that the inspector turns up each day and carries out the inspection as required by regulation 33(3):

> *Where the person owing a duty under paragraph (1) or (2) is an employee or works under the control of another, his employer or, as the case may be, the person under whose control he works shall ensure that he performs the duty.*

The reports prepared by the inspector should be kept in accordance with regulation 33(4) as follows:

> *The person on whose behalf the inspection was carried out shall –*
>
> *(a) keep the report or a copy of it available for inspection by an inspector appointed under section 19 of the Health and Safety at Work etc. Act 1974 –*
>
> > *(i) at the site of the place of work in respect of which the inspection was carried out until that work was completed, and*
> >
> > *(ii) after that for 3 months,*
>
> *and send to the inspector such extracts from or copies of it as the inspector may from time to time require.*

The requirements for keeping the report are self-explanatory. However, to keep the reports for longer than 3 months after completion of the work would be prudent as part of the record of the project history which may become relevant for any later contract disputes.

Specific hazards

Demolition or dismantling

In the 1994 Regulations, demolition work was given special priority on the basis that there was no exemption from the 1994 Regulations for demolition work for criteria that would otherwise indicate it was non-notifiable. In the Regulations there is no special treatment or consideration for demolition and as such is treated like any other construction work which has to be assessed on its own merits for hazards and associated risks.

Regulation 29 makes provision for demolition or dismantling as part of the consideration for safety on construction sites as follows:

> *(1) The demolition or dismantling of a structure, or part of a structure, shall be planned and carried out in such a manner as to prevent danger, or where it is not practicable to prevent it, to reduce danger to as low a level as is reasonably practicable.*

> *(2) The arrangements for carrying out such demolition or dismantling shall be recorded in writing before the demolition or dismantling work begins.*

The requirement for recording in writing arrangements for demolition or dismantling will apply to all construction work regardless of whether or not it is non-notifiable. There is no indication as to the detail that will be required to satisfy regulation 29(2), although for all intents and purposes it will be a risk assessment and method statement which is analogous to a construction phase plan in the case of non-notifiable projects.

Where the project is notifiable, the requirements to comply with regulation 29(2) are in addition to the requirements for a construction phase plan. The written arrangements can be incorporated in the construction phase plan or there is no reason why they cannot exist as a separate document.

Explosives

Merely the storage and transport of explosives is a hazardous operation, which is recognised by regulation 30(1) as follows:

> *So far as is reasonably practicable, explosives shall be stored, transported and used safely and securely.*

Explosives may be used in demolition, excavation or tunnelling. An excavation or a tunnel are both places of work that are required to be safe within the meaning of regulation 26. The requirement to ensure that no person is at danger of harm when explosives are detonated is set out in regulation 30(2) as follows:

> *Without prejudice to paragraph (1), an explosive charge shall be used or fired only if suitable and sufficient steps have been taken to ensure that no person is exposed to risk of injury from the explosion or from projected or flying material caused thereby.*

As part of the planning in the use of explosives, the contractor should pay due regard to maintaining a safe place of work, emergency routes and procedures, the instability of structures caused by vibration, the views of the competent person appointed to carry out inspections of excavations, cofferdams and caissons, and the prevention of risk from fire, fire detection and fire fighting.

Energy distribution installations

Electric power cables are a particular risk, either because of the interference with overhead power cables with on-site plant and equipment or cutting through buried cables. The first challenge is to locate such installations as required by regulation 34(1) as follows:

> *Where necessary to prevent danger, energy distribution installations shall be suitably located, checked and clearly indicated.*

To deal specifically with the risk from electric power cables, regulation 34 provides that:

> *Where there is a risk from electric power cables –*
>
> *(a) they shall be directed away from the area of risk; or*
>
> *(b) the power shall be isolated and, where necessary, earthed; or*
>
> *(c) if it is not reasonably practicable to comply with paragraph (a) or (b), suitable warning notices and –*
>
> > *(i) barriers suitable for including work equipment which is not needed, or*
> >
> > *(ii) where vehicles need to pass beneath the cables, suspended protections, or*
> >
> > *(iii) in either case, measures providing an equivalent level of safety, shall be provided or (in the case of measures) taken.*

The cost associated with relaying cabling or isolating power can be considerable and should be identified as a necessary step in the project programme, thus requiring the allocation of time and other resources.

The Regulations highlight the importance of this particular hazard by virtue of regulation 34(3) which requires that:

> *No construction work which is liable to create a risk to health or safety from an underground service, or damage to or disturbance of it, shall be carried out unless suitable and sufficient steps (including any steps required by this regulation) have been taken to prevent such risk, so far as is reasonably practicable.*

Prevention of drowning

Construction workers who have to work over water, on the water, or below the water level are at risk of drowning if they fall into the water or the water inundates them.

Construction workers are at risk of drowning in liquids other than water. Liquids are found in the food and drink manufacturing sector and other industrial processes. The duty to prevent drowning is in regulation 35(1), which provides that:

> *Where in the course of construction work any person is liable to fall into water or other liquid with a risk of drowning, suitable and sufficient steps shall be taken –*
>
> *(a) to prevent, so far as is reasonably practicable, such person from so falling;*
>
> *(b) to minimise the risk of drowning in the event of such a fall; and*
>
> *(c) to ensure that suitable rescue equipment is provided, maintained, and when necessary used so that such person may be promptly rescued in the event of such a fall.*

The preventative measures that should be considered for adoption as a risk prevention measure can include barriers, non-slip surfaces, harnesses and safety nets.

Minimising the risk of drowning can include draining the water or other liquid away or to such a depth that the risk of drowning is minimised.

Equally, persons who are required to work with the risk of drowning should, under certain circumstances, be issued with personal protective equipment, including life jackets, that will ensure that they remain buoyant if they fall into the water or other liquid. Wherever water presents a hazard to construction workers or any other person, there should be suitable rescue equipment that will include buoyancy aids, resuscitation equipment, ropes and, if necessary, small rescue craft.

Where construction workers have to be transported to the construction site across water, regulation 35(2) applies as follows:

> *Suitable and sufficient steps shall be taken to ensure the safe transport of any person conveyed by water to or from any place of work.*

The transport of workers by water should pay particular regard to the transfer from the land to the craft and from the craft onto the construction site.

A boat that is used to convey construction workers cannot be safe if it is overcrowded or overloaded as referred to in regulation 35(3), which provides that:

> *Any vessel used to convey any person by water to or from a place of work shall not be overcrowded or overloaded.*

Fire

Regulation 38 requires that:

> *Suitable and sufficient steps shall be taken to prevent, so far as is reasonably practicable, the risk of injury to any person during the carrying out of construction work arising from –*
>
> *(a) fire or explosion;*
>
> *(b) flooding; or*
>
> *(c) any substance liable to cause asphyxiation.*

Apart from the obvious risk associated with the use of explosives, which is subject to regulation 30, the use of fire or any other naked flame is a potential hazard. Flooding is a hazard that should be assessed and taken into account as part of the design and construction process. Substances liable to cause asphyxiation will include gases such as carbon monoxide or methane, or particulate materials, including powders and certain foodstuffs, due to inundation.

There is always a risk of fire and the sooner fires are detected the better the opportunity to contain the fire and mitigate its consequences. Regulation 41(1) refers to fire-fighting equipment and fire detection as follows:

> *Where necessary in the interest of health and safety of any person at work on a construction site there shall be provided suitable and sufficient –*

(a) fire fighting equipment; and

(b) fire detection and alarm systems,

which shall be suitably located.

The enforcing authority in respect of this duty will be the relevant fire authority.

In planning for the type of fire-fighting equipment and detection systems and locations where they should be provided on a construction site, the contractor should take account of the points under regulation 39(2) as required by regulation 41(2) as follows:

> *In making provision under paragraph (1) account shall be taken of the matters in regulation 39(2).*

The nature and extent of the fire-fighting equipment will be dependent upon the nature of the construction site, the work equipment, the number of persons working on a construction site and the physical and chemical properties of other materials on the site. A full risk assessment will be necessary to ascertain what fire-fighting equipment will be necessary.

The fire-fighting equipment and detection systems should be regularly tested as required by regulation 41(3) as follows:

> *Any fire fighting equipment and fire detection and alarm system provided under paragraph (1) shall be examined and tested at suitable intervals and properly maintained.*

Fire-fighting equipment that is not automatic should be clearly identified and easily accessible as required by regulations 41(4) and 41(7) as follows:

> *(4) Any fire fighting equipment which is not designed to come into use automatically shall be easily accessible.*

> *(7) Fire fighting equipment shall be indicated by suitable signs.*

Fire-fighting equipment requires training and there may be particular construction activities that create a particular risk of fire. In these circumstances, regulations 41(5) and 41(6) provide that:

> *(5) Every person at work on a construction site shall, so far as is reasonably practicable, be instructed in the correct use of any fire fighting equipment which it may be necessary for him to use.*

> *(6) Where a work activity may give rise to a particular risk of fire, a person shall not carry out such work unless he is suitably instructed.*

7 Competence

Introduction

The concept of competence underpins the Regulations because no duty holder can be appointed unless they are competent as required by regulation 4(1)(a), as follows:

> *No person on whom these Regulations place a duty shall appoint or engage a CDM co-ordinator, designer, principal contractor or contractor*

unless he has taken reasonable steps to ensure that the person to be appointed or engaged is competent.

The person so appointed or engaged also has an obligation to assess his or her own competence to avoid the risk of accepting an appointment or engagement if he or she considers him- or herself not to be competent. This duty is succinctly set out in regulation 4(1)(b), as follows:

No person on whom these Regulations place a duty shall accept such an appointment or engagement unless he is competent.

The duty to ensure that all persons on a project are competent is wider than persons appointed or engaged, and extends to arranging or instructing workers to carry out or manage design or construction work as required by regulation 4(1)(c):

No person on whom these Regulations place a duty shall arrange for or instruct a worker to carry out or manage design or construction work unless the worker is –

(i) competent; or

(ii) under the supervision of a competent person.

Since it is very unlikely that all workers on a project are competent, paragraph (ii) requires that they must be supervised by a competent person. Part 4 of the Regulations specifically requires a competent person for the inspection of excavations and cofferdams and caissons in accordance with regulations 31 and 32 respectively.

What is competence?

'Competent' is not defined in the Regulations and therefore persons who are required to assess competence have to rely on the ordinary definition of competent and the guidance provided in the ACOP.

It may be easy to rest back in the knowledge that a person who is competent can be recognised in the same way as an elephant can be easily recognised although difficult to describe. Unfortunately, for any person to adopt such a laissez-faire approach to the assessment of competence will be in breach of the Regulations.

The *Oxford English Dictionary* (2006) definition of the relevant aspects of competent is:

(1) suitable, appropriate;

(2) sufficient or adequate in amount, extent or degree;

(3) legally authorised or qualified, able to take cognizance (of a witness, evidence), eligible, admissible;

(4) having adequate skill, properly qualified, effective.

The definition identifies that competence is based on being suitable and appropriate to undertake a task successfully. The need for suitability and appropriateness is only to the extent that the degree of competence is sufficient or adequate. Skill and/or knowledge can be obtained by formal training qualifications or experience but the standard to which the task should be performed by a competent person has only to be sufficient or adequate. Competence is therefore to be assessed with respect to the particular task and the associated hazards and risks. The assessment of competence should not be compared against a 'gold standard' but related to that which is necessary to achieve a satisfactory result.

The principle of adequacy is reflected in regulation 4(2) by limiting the scope of competence as follows:

Any reference in this regulation to a person being competent shall extend only to his being competent to –

(a) perform any requirement; and

(b) avoid contravening any prohibition

imposed on him by or under any of the relevant statutory provisions.

In addition to the dictionary definition of 'competent', the meaning of competent has been examined within the context of health and safety legislation by the courts, whose decisions help to further an understanding as to where the boundary is to assess satisfactory or adequate skills and knowledge.

Competence as referred to in the Construction (Working Places) Regulations 1966 was applied to persons responsible for erecting, dismantling and installing scaffolds. A competent person was required to be responsible for inspection and supervision of competent workmen, 'possessing adequate experience of such work'. In the case of *Maloney* v. *A. Cameron Ltd 1961* 2 All ER 934, the Court of Appeal was prepared, without further comment, to recognise the competence of three painters for the purposes of erecting trestles in accordance with the Building (Safety, Health and Welfare) Regulations 1948.

The Quarries Order 1956 required the manager of a quarry to make and ensure the efficient carrying out of arrangements to secure that every inspection is carried out or done by a 'competent person'. In the absence of any definition of 'competent', Winn J in *Brazier* v. *Skipton Rock Co. Ltd 1962* said:

> *I am not prepared to hold either that 'competent' means the most competent person available to the owners of the quarry or the manager, or that it means that he shall be so competent that he never makes a mistake. In my judgment, it means a man who, on a fair assessment of the requirements of the task, the factors involved, the problems to be studied and the degree of risk danger implicit, can fairly, as well as reasonably, be regarded by the manager, and in fact is regarded at the time by the manager, as competent to perform such an inspection.*

The judge went on to say:

> *... experience is so often of much greater value than book-learning.*

A later decision by Cantley J in *Gibson* v. *Skibs A/S Marina 1966* in relation to the Dock Regulations 1934 elaborated:

> *Who is 'a competent person' for the purpose of such an inspection? This phrase is not defined. I think that it is obviously to be taken to have its ordinary meaning of a person who is competent for the task. I think that a competent person for this task is a person who is a practical and reasonable man, who knows what to look for and knows how to recognise it when he sees it.*

Competence as a requirement for persons fulfilling roles in health and safety legislation is now well established and found in many statutory provisions. The roles under health and safety legislation, where competence is either an express or implied requirement, have been mapped in the Executive's publication *An Outline Map on Competence, Training and Certification* prepared by the Policy Group and available on the Executive's website. In many examples to be found in health and safety legislation, competence is measured by a person having been issued with a licence to deal with a particular matter. In those circumstances it is easy to identify whether a person is competent merely by seeking evidence of the appropriate licence.

The challenge for any person assessing competence under the Regulations is to ensure that they have taken reasonable steps to assess competence to a point where they are satisfied, not just reasonably satisfied. The ACOP refers to a two-stage test. The person undertaking the assessment is required to focus on the needs of the particular project and is reminded that the assessments should be proportionate to the risks, size and complexity of the work. Inevitably, there is a considerable margin for variation between the extent, scope and depth of an assessment depending upon the judgement of the person making the assessment. The ACOP refers to the issue of judgement in the following terms:

> *If your judgement is reasonable, and clearly based on the evidence you have asked for and have been provided with, you will not be criticised if the company you appoint subsequently proves not to be competent when carrying out the work.*

For many persons this will provide little comfort. What is, or is not, reasonable is a question of fact to be decided ultimately by a court. Making the judgement as to what is reasonable will depend on the competence of the person making the assessment. There is no requirement upon a client to be competent! It is uncertain whether the appropriate test to be applied will be objective or subjective, although in the two cases referred to above the Court applied an objective test. It is reasonable to assume, therefore, that when the Courts are asked to determine questions of fact they will adopt an objective approach and, in the Magistrates Court, err towards the 'gold standard' rather than what might merely have been necessary.

The person undertaking the assessment is assisted in proving that they have taken reasonable steps to assess competence if they have followed the guidance in Appendix 4 to the ACOP, which is set out in full at Appendix 4.

Each step is made up of three elements. The first element in column 1 is the particular criteria to be assessed. The criteria have to be assessed against a standard to be achieved, which is described in column 2, and, finally, there are examples given of the evidence in column 3, which will be acceptable to support the assessment of competence.

The ACOP provides a word of warning, as follows:

> *Unnecessary bureaucracy associated with competency assessment obscures the real issues and diverts effort away from them.*

Despite the Executive's desire to have made the bureaucratic and administrative workload of the 1994 Regulations more reasonable, compliance with the ACOP assessing competence is likely to have negated such improvements elsewhere within the Regulations.

Although a person can have been assessed as competent before commencing construction work, the client is under a continuing duty to keep competency under review by virtue of regulation 9(2). Therefore, there would be no defence to a person having assessed competence having used reasonable judgement and taken the appropriate steps if the company that had been appointed proves not to be competent but then takes no further action when put on notice that the company is not so competent.

Achievement of competence

Competence is related to the particular project or task that the person being assessed will be undertaking. The ACOP encourages a focused approach to assessment of competence and reminds the assessor that the rigour of such an assessment should be proportionate to the risks arising from the construction work.

The assessments for a project will include assessing organisations and individuals. It follows that both organisations and individuals must attain the appropriate level of competence for the particular job.

An organisation cannot be competent unless it has competent individuals with the responsibility for the relevant aspects of the project. As a 'top down' approach an organisation will achieve competence by putting in place a management structure and procedures that are adequate to deal with the health and safety issues arising on a project. Conversely, on a 'bottom up' approach the individuals will have to demonstrate that they have acquired a particular level of education, training and experience.

The core criteria for demonstrating competence, provides the checklist for an assessor and also provide for the organisation or individual to be assessed with a clear indication of what is required.

The labourers and tradesmen working on a construction site do not need to be assessed for competence if they are being supervised by a competent person. However, these workers should be briefed as to the site rules and particular site risks in accordance with the contractor's duties. Within the design office, junior designers should also be supervised by competent persons who understand the site hazards and the risks which need to be avoided.

For any individual to be competent it requires a knowledge and understanding of the health and safety legislation, which will include regulations for specific risks in addition to the Regulations. It will also require demonstrable skill, expertise and experience. There will be many instances where the risks are low and the demonstration of competence, being assessed as proportionate to those risks, will be relatively easy to achieve. However, while many persons will consider that health and safety management is based on common sense, it is certainly not intuitive and formal training will be necessary for most persons on most projects.

Training can be provided through formal lecture/classroom teaching and by on-the-job monitoring and assessment, backed up by toolbox talks and short classroom sessions.

The point at which a person moves to the status of competence is not defined and is a matter of judgement for the assessor. In certain cases, the person's competence will depend upon obtaining a licence or statutory status without which the person cannot be assessed as competent.

The ACOP provides guidance as to the assessment of competence and training, which is a good starting-point. The training market is competitive and organisations and various institutions will be seeking to have their particular training recognised as a benchmark for competence. The assessor should not assume that an organisation that has provided the training is itself competent without some enquiry. Equally, an assessor should not rely entirely on certificates of training and education attainment if the risks are very serious and part of a complex project.

Criteria for demonstration of competence: Stage 1

The ACOP recommends a two-stage approach to the assessment of competence which has the benefit of providing assessors with a framework for the assessment. The assessor should not, however, treat the core criteria for the demonstration of competence as a perfunctory checklist.

The assessor should keep the assessment focused upon the nature of the project-specific tasks, the seriousness of the risks and overall complexity of the project. The assessment should be proportionate to the overall assessment of the risks to which construction workers are exposed. This is a question of judgement for the assessor which the assessor should record in writing in the event of any later enquiry. The assessment procedure should be recorded in writing and monitored throughout the project. The assessor should never assume that an assessment of competence at the beginning of a project will be maintained by an organisation or individual through to completion of a project.

1. *Health and safety policy and organisation for health and safety*
    The standard to be achieved is compliance with section 2(3) of the Health and Safety at Work etc. Act which states:

    > *Except in such cases as may be prescribed, it shall be the duty of every employer to prepare and often as may be appropriate revise a written statement of his general policy with respect to the health and safety of his employees and the organisation and arrangements for the time*

being in force for carrying out that policy, and to bring the statement and any revision of it to the notice of all his employees.

The evidence that is required by the ACOP is a signed, current copy of the company policy with indication as to when it was last reviewed and by whose authority it was published.

2.   *Arrangements*
The standard to be achieved is a clear explanation of arrangements for health and safety generally. This will include organisational arrangements. Arrangements for complying with the Regulations form only one part of the standard to be achieved.

A company is required to produce evidence of the arrangements for health and safety management generally and in particular discharging its duties under the Regulations.

3.   *Competent advice – corporate and construction related*
All organisations working within the construction industry must have access to competent health and safety advice. The ACOP expresses a preference for such advice being available within the organisation, although this should not be necessary if an organisation has appropriate arrangements with external consultants.

The health and safety adviser should be capable of providing general health and safety advice in addition to specialist construction industry advice and in particular the Regulations.

Evidence of reaching the standard for these criteria should include the name and competency details of the source of advice. The ACOP recognises that this can include a health and safety specialist team, trade federation or consultant. The ACOP also refers to the need for an example of the last 12 months of advice given and action taken.

4.   *Training and information*
A competent organisation should have in place and implement training arrangements, such that employees can discharge their duties under the Regulations. The training should include initial instruction and refresher training. Importantly, there should be a training programme throughout an organisation from board level to trainees.

The evidence for effective training will include training records, certificates of attendance, health and safety induction training for site-based workers and a programme for future training.

5. *Individual qualifications and experience*

   Employees are expected to have qualifications and experience appropriate for the tasks that will be undertaken by the persons on the project, unless they are under controlled and competent supervision.

   The qualifications and experience will vary depending upon the specified role under the Regulations, the specific tasks to be undertaken by the individuals and the size, complexity and the technical content of the project. The qualifications and experience for contractors, design organisations and CDM co-ordinators are set out in Appendix 4.

6. *Monitoring, audit and review*

   A competent organisation should have a system for monitoring its procedures that are audited at periodic intervals. A competent organisation should also review such procedures on an ongoing basis.

   The internal notes of meetings and correspondence connected with a formal audit and discussions with senior managers would be required as evidence. Evidence of the reports by the health and safety advisers following site visits or investigating accidents would also be good evidence.

7. *Workforce involvement*

   A competent organisation should have an established means of consulting with its workforce on health and safety matters.

   The ACOP suggested that the records of health and safety committees and names of appointed safety representatives will be good evidence. A consultation can be carried out in many ways including annual interviews with employees and suggestion boxes, in addition to meetings convened specifically to discuss health and safety.

   For organisations that employ less than five persons, they should be able to produce evidence by way of describing how they consult with their employees.

8. *Accident reporting and enforcement action; follow-up investigation*

   Any competent organisation should have records of all **RIDDOR** reportable events for at least 3 years. An organisation should also have in place a system for reviewing all incidents, recording the action taken following the incident.

   An organisation should also keep a record of its dealings with the Executive with respect to enforcement action including improvement notices, prohibition notices and prosecutions.

A company should be able to produce the documentary records associated with accident reporting and enforcement action. Large organisations should be able to produce statistics. Importantly, an organisation should be able to produce evidence of the methodology for recording and investigating accidents and incidents.

9.   *Sub-contractor/consulting procedures*

This criterion only applies if a person intends to sub-contract the design or construction work. In the event a person intends to appoint sub-contractors, he or she has to have arrangements in place to ensure that the appointed sub-contractors are competent. The appointor will be expected to demonstrate how they will ensure that sub-contractors will also have arrangements for appointing competent sub-contractors or consultants. The arrangements should also include the means for monitoring sub-contractor and consultant performance.

The supporting evidence is not required to be prescriptive and can be limited to show how the person will ensure that sub-contractors or consultants are competent. This can be provided by previous examples of sub-contractor assessments which have been carried out.

The ACOP also refers to the need for evidence showing how similar standards of competence assessment are achieved from different sub-contractors.

As compared between sub-contractors undertaking different tasks under different disciplines, this evidence is likely to cause particular difficulty. Finally, a person needs to show how sub-contractor performance will be monitored with respect to health and safety and the Regulations.

10.   *Hazard elimination and risk control*

The ACOP identifies designers only as being subject to these criteria. The standard to be achieved by designers is that required to have been implemented for arrangements for meeting the duties of designers under regulation 11.

Evidence to demonstrate meeting the required standard will involve co-operation and co-ordination of design work within the design team and with other designers, and ensuring that hazards are eliminated and any remaining risks are controlled. It is also necessary to demonstrate how any structure that will be used as a workplace will meet the relevant requirements of the Workplace (Health, Safety and Welfare) Regulations 1992.

Examples from previous projects showing how risks were reduced through design will be supportive of the other evidence.

The ACOP also refers to the evidence of a short summary as to how changes to designs were managed. This should include the important information as to how design changes will be communicated to other persons. The ACOP points to a preference for practical measures that reduce particular risks as opposed to lengthy procedural documentation on generic risks. The safe approach will also be to produce evidence for all risks both generic and particular, as to do otherwise is to run the risk that somebody else may take a different view of a generic risk.

11. *Risk assessment leading to a safe method of work*

These criteria are applicable to contractors only who are carrying out construction work. A competent contractor should have procedures in place for carrying out risk assessments and for developing and implementing safe systems of work and method statements. The ACOP draws attention to the importance of identifying health issues.

The evidence that the contractor will need to produce will identify how he or she identifies significant health and safety risks and how these will be controlled. At the very least, there should be some risk assessments, method statements and examples of safe systems of work.

Contractors who employ less than five persons and do not have written arrangements can provide such evidence orally.

The ACOP draws attention to the need for evidence that reflects the importance of the risks and the whole approach to risk assessments.

12. *Co-operating with others and co-ordinating your work with that of other contractors*

The contractor should be able to illustrate how co-operation and co-ordination of his or her work is achieved in practice, and how the workforce is involved in drawing up method statements and safe systems of work.

The evidence should include sample risk assessments, procedural arrangements and project team meeting notes. Evidence will also be required of how the company co-ordinates its work with other contractors.

13. *Welfare provisions*
     The standard for welfare provisions should comply with Schedule 2
     of the Regulations and in particular that such provisions will be in
     place before people start work on site.
         Evidence would include health and safety policy, commitment,
     contracts with welfare facility providers and details of type of
     welfare facilities provided on previous projects.

14. *CDM co-ordinator's duties*
     The CDM co-ordinator should be able to demonstrate how he or
     she will go about encouraging co-operation, co-ordination and
     communication between designers.

The ACOP indicates that the evidence should be in the form of actual
samples rather than by generic procedures. Appendix 5 of the ACOP, as
set out in Appendix 4 in this book, provides guidance for assessing
competence of a CDM co-ordinator for large or complex projects or
for projects with high or unusual risks.

Criteria for demonstration of competence: Stage 2

1.   *Work experience*
     The person to be assessed should give details of relevant experience
     in the field of work for which he or she is being assessed. Inevitably,
     the evidence for the Stage 2 assessment will be more project focused
     and include health and safety records from previous projects, and
     names of persons able to provide references. In respect of projects
     that are materially different in type, complexity or scope to the
     project in question, the person to be assessed should be able to
     identify the shortcomings in their experience and how they
     intend to overcome such shortcomings.

8 The client

Introduction

It is unarguable that the role of the client is central to the procurement of construction work. This is due to the client's influence over the budget and other resources and the time frame within which the project is conceived through to completion of construction. The Regulations have recognised the practicalities of real projects and have enhanced the role of the client and ensured that clients are obliged to take a positive management role and can no longer contract out of their health and safety responsibilities by appointing agents for a project.

The challenge for the construction professionals is to ensure that all clients, from sole traders to the largest multinational companies, are made aware of the general health and safety issues, such that embracing the Regulations is seen as a positive step instead of a bureaucratic chore. Meeting the challenge should be seen as very much part of the health and safety training of construction professionals in preaching the virtues of health and safety management to the huge number of cost-conscious smaller clients whose awareness is considerably less than the larger corporate client organisations.

The definition of client is set out in regulation 2(1) as follows:

> *Client means a person who in the course or furtherance of a business –*
>
> *(a) seeks or accepts the services of another which may be used in the carrying out of a project for him; or*
>
> *(b) carries out a project himself.*

By virtue of the Interpretation Act 1978, 'person' includes corporations and limited companies as well as individuals. Therefore, anybody or any organisation embarking on a project, that includes or is intended to include construction work which itself includes all planning, design, management or other work involved in a project until the end of the construction phase, is subject to the Regulations. However, identifying the client may not be without some confusion as between landowners, investment purchasers, funders, insurers, tenants or developers. In circumstances where it may be unclear as to the identity of the client out of several possible choices, regulation 8, which is discussed below, allows one or more clients to make an election that they are to be treated as the client for the purposes of the Regulations.

A person becomes a client merely if they seek to accept the services of another who may be used in the carrying out of a project. Therefore, at the very outset, when inviting tenders from consultants or contractors, before there may be any settled decision to embark upon construction work, the person considering such appointments is subject to the duties of the client created by the Regulations. A client may consider carrying out part of a project him- or herself, particularly the initial stages of design. Even in these circumstances, the person is bound by the duties of a client pursuant to the Regulations. Thus, at the outset, a person who will be acting as client has to take steps to assess his or her own competence to undertake those tasks as part of a project.

Client types

Domestic clients

The definition of client requires that the person is acting in the course of, or furthering, a business.

A 'business' means:

> *a trade, business or other undertaking (whether for profit or not).*

The concept of a domestic client is not addressed specifically by the Regulations but is a person acting as a client not in the course of furtherance of a business. By exception, a client who wants to be excluded from the Regulations has to demonstrate that procuring the services of another as part of the carrying out of a project (regulation 2(1)(a)), or undertaking the construction work itself (regulation 2(1)(b)) is not in 'the course or furtherance of a business'. While this can clearly be attributed to construction work on domestic dwellings, where the dwelling is the principal residence of the 'client', the exception is unlikely to apply to the building of another dwelling in the curtilage of the principal residence, or the conversion of the family house into apartments. If in doubt, a reasonable test is to ascertain whether the fruits of the building activity would lead to a tax liability on any profit, in which case the likelihood is that the domestic client has created a 'business' and therefore comes within the scope of the Regulations, including notification, if the project is likely to last longer than 30 days.

A designer or contractor appointed by a client who is not in the course or furtherance of a business has no duty to make the client aware of the Regulations. However, it would be unwise for a designer or a contractor to assume a project was merely domestic and should make appropriate enquiries of a client before commencing work. Just because a designer or contractor is working for a domestic client does not relieve them of their duties under the Regulations.

The Crown as client

Her Majesty's government acts in the name of the Crown, which includes all government departments and armed forces.

By virtue of section 48 of HASWA 1974, the Crown is subject to the statutory duties set out in the Act. However, it cannot be the subject of enforcement action by the Executive (sections 21–25), whether by way of improvement or prohibition notices or by way of prosecution (sections 33–62). Crown censures are formal recordings of a decision by the Executive that, but for Crown immunity, the evidence of a

Crown body's failure to comply with health and safety law would have been sufficient to provide a realistic prospect of conviction in the courts. Accordingly, those government departments and organisations that come within the definition of the Crown have a duty to observe the provisions of the Act and the regulations made under it, including the Regulations. This is as it should be, since the declaration by Construction Minister, Viscount Ullswater in 1994, in response to the publication of the Latham Report, stated that:

> *The government aspires to being a best practice client, and I am happy to make that declaration.* (Department of Environment press release 25th July 1994)

Employees of the Crown can, however, be prosecuted by virtue of section 48(2) of the Act, which is probably a sufficient incentive to ensure compliance with health and safety legislation.

In common with many other clients, the Crown is likely at times to fulfil other roles identified under the Regulations, in which case the same immunity from prosecution for the Crown and personal criminal liability of its employees will apply.

Local authorities and public bodies as clients

Local authorities, non-government organisations and health trusts within the National Health Service have no immunity from prosecution, unlike the Crown and the government departments which act in its name.

Insurance and warranty claims

Construction work that is procured as a result of a claim under an insurance policy may result in the insurance company being treated as the client. However, insurance companies are not experts in the construction process or resourced to undertake the management of construction projects. The loss adjusters that assess the need for construction work and become involved in procuring the work on behalf of an insurance company can be treated as the clients. In fulfilling such a role, agents of the insurance company should ensure that their professional indemnity insurance will provide sufficient and appropriate cover.

If the insured procures the construction work on the basis that the costs of construction are to be reimbursed by the insurance company, the insured will be treated as the client.

To avoid any doubt as to the person who will be treated as the client, which could include the insurance company, loss adjuster or other agent and the insured, it would be prudent for the person who has the appropriate expertise or interest as client to make the election in accordance with regulation 8.

Developers

In circumstances where a person buys a house before the construction is complete, it does not mean that the developer who has sold the land or the partially constructed house is no longer the client. The purchaser may have an interest in the property and may have specified certain features in the remainder of the construction phase but it is still the developer who arranges for the construction work to be done and they will be treated as the client.

Developers often fulfil numerous roles concurrently including client, designer, principal contractor and CDM co-ordinator. There is no prohibition on a developer fulfilling all the roles of duty holder as long as they take steps to ensure they fulfil all the requirements of each type of duty holder under the Regulations.

Private Finance Initiative, Public Private Partnership and similar forms of procurement

Public projects are frequently funded by private finance or a combination of private and public funds. It is a feature of such projects that the person fulfilling the role of client will evolve to reflect the different stages in the procurement process.

The project originator, which is likely to be a public authority, will in the first instance be the client. The role of client will move when the contract arrangements for undertaking the construction work have been resolved which is likely to be a special-purpose vehicle (SPV). The SPV may itself comprise separate organisations responsible for the construction whereas the other part of the SPV takes responsibility for the operational management of the project after construction. It is important for the project originator to recognise the consequences of remaining in the role of client if an SPV or other organisation has been formed especially for the procurement of design and execution of construction work.

The potential clients should make the decision as to which of them should make the election to be the client.

Election by clients

Clients are no longer able to appoint an agent to act as the only client for a project, which may or may not be notifiable, as was permissible under the 1994 Regulations. However, in the situation where there is a project with more than one client, the clients can make an election, in accordance with regulation 8 as follows:

> *Where there is more than one client in relation to a project, if one or more of such clients elect in writing to be treated for the purposes of these Regulations as the only client or clients, no other client who has agreed in writing to such election shall be subject after such election and consent to any duty owed by a client under these Regulations save the duties in regulations 5(1)(b), 10(1), 15 and 17(1) insofar as those duties relate to information in his possession.*

It is often the case that a developer on behalf of a purchaser or tenant procures the construction of a building. Within the definition of 'client' the organisation for whom the developer is procuring the construction could also be construed as a client. More obviously, organisations may come together on a joint venture basis to procure a project in which case it is open to those organisations to elect one or more organisations as the client for the purposes of the Regulations. At the outset of a project all the initial parties should be alert to the identity of potential clients.

In circumstances where an election is not made in writing it is implied that the responsibility of the clients is shared. It is not specified whether the shared responsibility of the clients is joint and several or proportional based on a measured responsibility according to some measure of contribution.

The difficulties of a proportional assessment would involve a detailed assessment of contribution be it resources including financial, technical, manpower or legal status including 'remoteness' from the management of the construction work or legal interest in the construction site.

For an election to be effective it must be made in writing, although it is not a requirement of notification to the Executive. The election will be a simple statement within the legal documentation which will record and set out the respective roles, duties and obligations of the clients in relation to a project. However, if there is more than one person electing to fulfil the role of client, the Executive is likely to treat such persons as jointly and severally liable under the Regulations to ensure that all the client's duties are discharged.

An election, by definition, will mean that one or more clients who have not elected are avoiding the full effect of the Regulations. Nonetheless, it is not possible for a client seeking to rely on an election to avoid some of the duties in the Regulations, specifically 'regulations 5(1)(b), 10(1), 15 and 17(1) insofar as those duties relate to information in his possession'. These minimal duties ensure that a client cannot remain disinterested in a project, being subject at all times to the duty of co-operation and provision of information.

General duties

Arrangements for managing projects

The client's influence over small or large projects is determined at the outset by the client's need and requirements for the project. The client's decision making, as to the scope of the project, design brief, project delivery date, project budget and willingness to compromise on aesthetics, choice of materials and timescales, cannot be overestimated. Regulation 9 leaves the client in no doubt as to the central role he will fulfil at the start and duration of a project. Regulation 9 sets out his duties in relation to arrangements for management projects as follows:

> *Every client shall take reasonable steps to ensure that the arrangements made for managing the project (including the allocation of sufficient time and other resources) by persons with a duty under these Regulations (including the client himself) are suitable to ensure that –*
>
> *(a) the construction work can be carried out so far as is reasonably practicable without risk to the health and safety of any person;*
>
> *(b) the requirements of Schedule 2 are complied with in respect of any person carrying out the construction work; and*
>
> *(c) any structure designed for use as a workplace has been designed taking account of the provisions of the Workplace (Health, Safety and Welfare) Regulations 1992 which relate to the design of, and materials used in, the structure.*

The client is not required to manage the project, although it may choose to do so but must, under all circumstances, allocate sufficient time and other resources, mainly appropriate funding, so that the management arrangements meet the three key objectives in paragraphs (a) to (c).

The client, therefore, must appoint other persons to the extent that it cannot or does not wish to undertake all the roles of the duty holders, to fulfil the tasks of design and construction and the role of CDM co-ordinator in the case of notifiable projects. Thus, the client has the prime responsibility to ensure that such persons who are appointed have been assessed for competence and, in turn, that such persons will have the competence to assess those who they will be appointing. The 'pyramid of competence' relies on the fact that the client has taken an appropriate view of the particular hazards and risks of a project and assessed the competence of the persons he or she will appoint and by doing so set the tone for the entire project.

The client is unlikely to carry out construction work unless he or she has a direct labour capability so that the objective to have the work carried out, so far as is reasonably practicable, without risk to health and safety of any person, starts with the appointment of competent persons. The client should not impose unrealistic constraints upon the designers and the contractors, which will jeopardise health and safety management. In particular, the client should avoid imposing shorter timescales, undermining co-operation and co-ordination between the parties and ultimately failing to communicate the information required by regulation 10. Where construction work is planned to take place within existing premises which are currently being used as a workplace or on a site adjacent to the client's operations, the client should take steps to ensure that his or her own activities do not create any unforeseen risks.

The client should always ensure that the workers on a construction site have access to the welfare facilities, which are listed and described in Schedule 2 and more fully discussed in Chapter 5. The client does not have to possess any particular skills and experience to ascertain whether a contractor is proceeding in compliance with Schedule 2. If a client is in any doubt as to why certain facilities would not be made available, it is incumbent upon the client to satisfy him or herself that the provision of such welfare facilities would have represented an unreasonable requirement under all the circumstances. The client can assist contractors during the construction work in many instances by extending the use of his or her own welfare facilities for the use of construction workers. Indeed, this would be expected if the contractor was otherwise unable to provide such facilities when the client's own facilities are readily accessible.

The duty upon a client to ensure that a structure which will be used as a workplace has been designed in accordance with the provisions of the Workplace (Health, Safety and Welfare) Regulations may appear unreasonable if the client has no input into the design. In practice, however, a client procuring a project for a workplace for its own use is likely to have a considerable influence and contribution to the design brief. The designer, who will have been assessed as being competent by the client, has a duty to design such a structure in accordance with such provisions of the Workplace (Health, Safety and Welfare) Regulations by virtue of regulation 11(5). If the client has made a contribution to the design brief which is entirely unreasonable or unsafe from a health and safety management point of view, the designer should advise the client. In those circumstances, the client is not in a position to insist upon his or her own judgement. If the designer recognises that the client is unprepared to change the design brief or is interfering with the design so as to compromise the designer's own duty under the Regulations, the designer should terminate his or her appointment.

As a general proposition, any party to a project who has reasonable grounds to have concluded that the client is not properly exercising its duty with respect to the arrangements for managing projects may have grounds for terminating their contracts of appointment with the client.

The contribution to the management of a project by a client does not end with making appropriate and lawful appointments. Regulation 9(2) requires that:

> *The client shall take reasonable steps to ensure that the arrangements referred to in paragraph (1) are maintained and reviewed throughout the project.*

The client's primary duty, therefore, is to ensure that those persons appointed to the different roles for the duty holders maintain their competence throughout the project. The client should be alert to any developing shortcomings in the health and safety management by the duty holders and should have reserved rights within the contracts of appointment to deal with such circumstances.

In the case of notifiable projects, the client can rely upon the CDM co-ordinator to provide advice and assistance to ensure compliance with regulation 9. However, where small projects are concerned, which are not notifiable, the client has to rely upon his or her own judgement, after taking reasonable steps to ensure overall compliance with regulation 9. If a small project has particular dangers or high risks likely to cause harm to health or injury, the client should appoint a person to fulfil the role of a quasi CDM co-ordinator in much the same way as the ACOP recognises the need for a quasi construction phase plan for non-notifiable projects.

Provision of information

Risk assessment is only as good at preventing accidents as the expertise in recognising and identifying hazards. For many projects, the hazards would not be immediately apparent to experienced designers and contractors. Where hazards are intrinsically connected to the site or the existing structure, the person most likely to have the best state of knowledge will be the client. It is for this reason that the transfer of pre-construction information from the client to the designers and contractors is an important step in minimising risks during the construction phase. Regulation 10(1) requires that:

> *Every client shall ensure that –*
>
> *(a) every person designing the structure; and*
>
> *(b) every contractor who has been or may be appointed by the client,*
>
> *is promptly provided with pre-construction information in accordance with paragraph (2).*

Thus, the client is required to identify all of the designers and contractors that he has appointed or may appoint. The client should recognise that suppliers are also very often designers and information should be provided to them. Pre-construction information should be provided promptly so that initial designs can take account of any hazards. There is an obvious commercial advantage for the client in providing such information as it may avoid wasted design time that could result from duplication when later information comes to light.

It is unreasonable to assume that many clients will recognise particular hazards. Especially for non-notifiable projects where the requirement for a CDM co-ordinator and construction phase plan are not necessary, the client should be entitled to expect relevant advice from the designer. However, the regulations do not impose a duty on the designer to draw such matters to the attention of the client.

The scope of the pre-construction information is described by regulation 10(2) as follows:

> *The pre-construction information shall consist of all the information in the client's possession (or which is reasonably obtainable), including –*
>
> *(a) any information about or affecting the site or the construction work;*
>
> *(b) any information concerning the proposed use of the structure as a workplace;*
>
> *(c) the minimum amount of time before the construction phase which will be allowed to the contractors appointed by the client for planning and preparation for construction work; and*
>
> *(d) any information in any existing health and safety file,*

which is relevant to the person to whom the client provides it for the purposes specified in paragraph (3).

A list of the pre-construction information can be found at Appendix 2 of the ACOP and is reproduced here at Appendix 2.

The client has a judgement to make as to the extent and scope of the pre-construction information he is to provide. In the case of non-notifiable projects, the client is left to his or her own judgement, which would suggest that the safe option is to supply all information which can conceivably be identified. Unfortunately, this approach would lead to an inefficient and burdensome process for the designers and contractors. However, under all the circumstances, a client could not be criticised for being conservative in identifying all the information. In the case of notifiable projects, the client can accept advice from the CDM co-ordinator to be selective in producing pre-construction information which is relevant.

The information which the client is required to produce is not just information about the site or the construction work. The pre-construction information should also include information that will affect the site. Therefore, the natural features or activities remote from the site that might create a hazard should also be identified. This might typically include flooding from a river, a delimited speed limit on the main access road to the site, or storage of hazardous chemicals on an adjacent site.

The client who is procuring the project for the purpose of building a workplace or making amendments to an existing workplace should provide pre-construction information to the designers and contractors which would be expected in any competent design brief. Perhaps not so obviously, the client should provide detailed information about the staff profiles, shift patterns, weekend working, public access, etc.

The client should have provided a reasonable period of time between providing the pre-construction information to contractors, one or some of whom will be required to start construction work by a target date. If necessary, time should be allowed for detailed investigations and surveys to verify design assumptions and confirm design philosophies and construction methods.

The existing health and safety file is an important source of information for existing structures that have been compiled pursuant to the 1994 Regulations or the new Regulations. The ACOP suggests that where there are gaps in the information, particularly in respect of the presence of any asbestos, the client should undertake commissioning surveys. The client should not be concerned to undertake such surveys if it has the benefit of competent designers. The client might risk wasting money on an inappropriate survey without the valuable input from competent designers who would appreciate what further information is needed for the gaps in the information.

The pre-construction information should be provided to designers as part of the design brief and to contractors as part of the invitation to tender or in the information pack prior to negotiating a contract. The purposes for which the information will be used by the designers and contractors are set out in regulation 10(3) as follows:

> *The purposes referred to in paragraph (2) are:*
>
> *(a) to ensure so far as is reasonable practicable the health and safety of persons –*
>
>> *(i)   engaged in the construction work,*
>>
>> *(ii) liable to be affected by the way in which it is carried out, and*
>>
>> *(iii) who will use the structure as a workplace; and*
>
> *(b) without prejudice to sub-paragraph (a), to assist the persons to whom information is provided under this regulation –*
>
>> *(i)   to perform their duties under these Regulations, and*
>>
>> *(ii) to determine the resources referred to in regulation 9(1) which they are to allocate for managing the project.*

Unless the client is experienced and knowledgeable with respect to the information that is likely to be required by designers and contractors, the information which is provided may not meet all the relevant purposes. The designers and contractors should analyse the pre-construction information to satisfy themselves, at least at face value, that the information assists them in protecting construction workers, other persons who are likely to be affected by the construction work and users and occupants of the workplace. In the event that the information has any obvious gaps, the client can expect to be pressed for further information or clarification. The exchange of information and ensuing dialogue will be a rehearsal for the co-operation and co-ordination that is expected from all the parties throughout the project.

The approach to be adopted by the designers and contractors in analysing the pre-construction information is to consider its completeness so that their own duties can be fulfilled under the Regulations and they can plan the construction work with the necessary allocation of resources to meet the timetable for commencement of construction work.

Duties for notifiable projects

Appointment of a CDM co-ordinator

The client has additional duties for notifiable projects. This is only to be expected because notifiable projects, of longer duration or greater intensity than non-notifiable projects, will inherently present greater risks. The client is, in many ways, perhaps more able to manage the additional duties and avoid the risk of breaching the Regulations for notifiable projects because of the appointment of the CDM co-ordinator. The client's obligations in respect of appointment of a CDM co-ordinator are set out in regulation 14(1) as follows:

> *Where a project is notifiable, the client shall appoint a person ('the CDM co-ordinator') to perform the duties specified in regulations 20 and 21 as soon as is practicable after initial design work or other preparation for construction work has begun.*

If the client has assessed that the project will be notifiable, the CDM co-ordinator will assist him or her to carry out the client's duties, to co-ordinate and manage health and safety aspects of the design work and finally prepare the health and safety file. There is nothing to prevent the client, in the case of non-notifiable projects, to appoint a person to fulfil a role equivalent to the CDM co-ordinator and by means of the contract of appointment, require such person to provide the same services which the CDM co-ordinator would provide under the Regulations. This may in fact be advisable where the risks associated with a non-notifiable project can be identified as unusually high and, in line with the ACOP recommendation, a document similar to a construction phase plan might be required.

The appointment of the CDM co-ordinator should be at the earliest possible stage in the defining of the project. Initial design work will benefit from a consideration of health and safety management issues and can have the commercial advantage of discounting certain solutions before further design work is undertaken and paid for.

Other preparation for construction work will include activities that are not, of themselves, construction work. This might include, for example, site surveys, traffic counting and similar tasks associated with early design or actual construction work, but sufficiently in advance of the project as to constitute non-notifiable preliminary works, which might include erection of fencing, soil investigations and demolition.

As a pre-condition to appointing the CDM co-ordinator the client has to ensure that the person to be appointed is competent. The Stage 1 and Stage 2 criteria for assessment of competence should be adopted by the client before confirming an appointment. For a larger or more complex project, or a project with high unusual risks, deeper skills and experience are required from the CDM co-ordinator than might be satisfactory for routine projects.

Appendix 5 of the ACOP which is reproduced as Appendix 4 sets out a further two criteria for Stage 1 and further criteria for Stage 2. There is no guidance for the client as to what might constitute a large or more complex project, or one with high or unusual risks. Without being able to make such an assessment by reference to any guidelines other than the client's own subjective judgement, the ACOP guidance is likely to be adopted for the appointment of CDM co-ordinators for most projects.

The person who will be assessed as competent for large and complex projects is likely to be professionally qualified, which the ACOP suggests is the equivalent of chartered status. In addition to attainment of chartered status, the person should also have acquired specific health and safety qualifications supported by validated continuing professional development.

The Stage 2 criteria for assessing the CDM co-ordinator is arguably the most important of all the criteria. Does the CDM co-ordinator have relevant experience for the type of project planned by the client?

The client is reminded in the ACOP that assessments should focus on the needs of the particular job and should be proportionate to the risks arising from the work. This is the Executive's attempt to limit the paper and administrative workload that clients might generate in assessing competence. The client should be comforted by the fact that the CDM co-ordinator is under a duty not to accept the appointment if he does not consider himself competent to undertake the tasks for the particular project.

Appointment of a principal contractor

The principal contractor has extensive duties in relation to a notifiable project which will involve the principal contractor in the expenditure of considerable time and cost during the pre-construction phase. If the principal contractor is appointed at an early stage, the client will mitigate the overall burden of costs for tendering in the construction industry that are thrown away when contractors are not appointed.

Conversely, if the client embarks upon a long period for allowing potential principal contractors to prepare tenders and run with a short-list close up to the commencement of the construction phase, the cost burden for tendering becomes extremely high. Regulation 14(2) requires the client to appoint a principal contractor as soon as he knows the identity of the person who will be appointed as principal contractor. Regulation 14(2) directs the client as follows:

> *After appointing a CDM co-ordinator under paragraph (1), the client shall appoint a person ('the principal contractor') to perform the duties specified in regulations 22 to 24 as soon as is practicable after the client knows enough of the project to be able to select a suitable person for such appointment.*

In making such an appointment, the client will have already appointed a CDM co-ordinator who will be available to give advice on the selection of the principal contractor before appointment. The principal contractor will have to have been assessed for competence by the client which, ideally, should have been undertaken before inviting the principal contractor to tender. The client has a more difficult task in assessing 'project chemistry', that is to say, can the client assess how the team comprising him- or herself, the CDM co-ordinator and principal contractor will operate? Unfortunately, these are issues which cannot be reduced to bald text within Regulations or Codes of Practice. Nonetheless, the client should ensure that he is comfortable that the principal contractor has demonstrated an appropriate style and approach to co-operation together with the mandatory pre-condition of ensuring that the principal contractor is competent. The client should adopt the Stage 1 and Stage 2 criteria as set out in Appendix 4 to assess the principal contractor. Echoing the remarks for the assessment by the client of the CDM co-ordinator's competence, the client should avoid unnecessary bureaucracy and note that the principal contractor is also required to have assessed his competence in accordance with regulation 4(1)(b).

Duty to review the appointments

The client is not allowed to become complacent by the requirement to keep the appointments for the CDM co-ordinator and principal contractor under review. Regulation 14(3) requires the client to keep the appointments under review as follows:

> *The client shall ensure that appointments under paragraphs (1) and (2) are changed or renewed as necessary to ensure that there is at all times until the end of the construction phase a CDM co-ordinator and principal contractor.*

The client may have appointed a CDM co-ordinator under terms of engagement to terminate before the construction phase if there is a prospect that construction work may not commence. In those circumstances, the CDM co-ordinator's appointment will require renewing. The same situation may arise during the construction phase if a separate contractor is engaged to undertake site clearance and demolition work. In those circumstances, the demolition contractor will be appointed as the principal contractor, whereas the main construction contractor will be appointed later as the principal contractor. This may be the same contractor if he has been successful in winning the main works under the construction phase.

The client should keep the appointments under review, not least in the event that the level of competence for either duty holder falls below a level that is required for the later construction work.

The client should include in the terms and engagement for either duty holder a mechanism for transferring information to a new incoming duty holder to avoid any disruption in the role.

The appointments of the CDM co-ordinator and the principal contractor are required to have been made in writing to comply with regulation 14(5) as follows:

> *Any reference in this regulation is to appointment in writing.*

This is of particular importance because without an appointment in writing there is no appointment, which has significant consequences for the client.

Duty of the client in the absence of appointments

What if the client fails to appoint the CDM co-ordinator or the principal contractor at the time when another client may have made the appointment much earlier? What if there are no appointments in writing? Regulation 14(4) provides the answer as follows:

> *The client shall –*
>
> *(a) be deemed for the purposes of these Regulations, save paragraphs (1) and (2) and regulations 18(1) and 19(1) to have been appointed as the CDM co-ordinator or principal contractor, or both, for any period for which no person (including himself) has been so appointed; and*
>
> *(b) accordingly the subject to the duties imposed by regulations 20 and 21 on a CDM co-ordinator or, as the case may be, the duties imposed by regulations 22 to 24 on a principal contractor, or both sets of duties.*

Thus, the client can unwittingly be subject to the duties of the CDM co-ordinator and principal contractor until such time as they have been appointed. If the client intends to fulfil the role of CDM co-ordinator, or principal contractor or both, the client is not required to set out an appointment in writing, although it should record in writing an assessment for competence to demonstrate compliance with regulation 4.

To avoid the possibility of a technical breach of the Regulations which would put the client at risk of being subject to the duties of the CDM co-ordinator and principal contractor, the client should specifically state in the written appointments that the relevant persons are appointed to the role of CDM co-ordinator or principal contractor as appropriate.

The client should be particularly alert during the construction phase to any contractual event which might terminate the appointment of the CDM co-ordinator or principal contractor.

Duty to provide information

Regulation 15 requires the client to provide pre-construction information to the CDM co-ordinator and principal contractor as follows:

> *Where the project is notifiable, the client shall promptly provide the CDM co-ordinator with pre-construction information consisting of –*
>
> *(a) all the information described in regulation 10(2) to be provided to any person in pursuance of regulation 10(1);*
>
> *(b) any further information as described in regulation 10(2) in the client's possession (or which is reasonably obtainable) which is relevant to the CDM co-ordinator for the purposes specified in regulation 10(3), including the minimum amount of time before the construction phase which will be allowed to the principal contractor for planning and preparation of construction work.*

The nature and scope of the pre-construction information that the client is required to produce is no different when compared with the information that the client has to produce for non-notifiable projects. The client's task is easier because he no longer has to identify all those persons to whom the information should be provided. The client's duty is discharged by providing all the pre-construction information to the CDM co-ordinator who has the task of distributing that information as appropriate.

The minimum amount of time allowed to the principal contractor before the construction phase is due to start is an important parameter for the CDM co-ordinator who has to manage the necessary co-operation and co-ordination thus ensuring the design is completed and the construction phase plan can be completed before construction work starts.

Commencement of the construction phase

Until the commencement of the construction phase there is no actual risk of harm to any persons. It follows, therefore, that the start of the construction phase is a critical moment in the life of the project. Regulation 16 requires the client to take account of the principal contractor's preparations immediately prior to the start of construction work as follows:

> *Where the project is notifiable, the client shall ensure that the construction phase does not start unless –*
>
> *(a) the principal contractor has prepared a construction phase plan which complies with regulations 23(1)(a) and 23(2); and*
>
> *(b) he is satisfied that the requirements of regulation 22(1)(c) (provision of welfare facilities) will be complied with during the construction phase.*

This requirement on the client creates an important hold point. If the construction phase plan has not been prepared by the principal contractor and the welfare facilities are not compliant with Schedule 2, the construction phase should not start. The client's duty is to be satisfied in respect of paragraphs (a) and (b). It is not acceptable for the construction phase to start if the client is satisfied with respect to the construction phase plan but has not made enquiries or seen any evidence that the welfare facilities are in accordance with Schedule 2 and vice versa. In considering the construction phase plan, the client's duty extends to ensuring that such a plan has been prepared. There is no express obligation on the client to consider the content of the construction phase plan or assess its accuracy, practicability or adequacy. However, this obligation can be implied by reference to regulation 9(1)(a) which requires the client to take reasonable steps to ensure that the arrangements for a project are suitable to ensure that the construction work can be carried out so far as is reasonably practicable without risk to the health and safety of any person. The client is entitled to rely upon the CDM co-ordinator to provide sufficient advice and assistance with respect to an assessment of the principal contractor's preparation of the construction phase plan. The client's duty is not subject to a test of reasonableness in regulation 16(a) other than the steps which the client takes to ensure the arrangements for managing the project shall be reasonable in accordance with regulation 9(1).

The provision of welfare facilities which the principal contractor is bound to provide pursuant to regulation 22(1)(c) does not have to be set out in the construction phase plan, although the principal contractor should have arrangements for the welfare facilities in an easily identifiable document describing the provision of welfare facilities to satisfy the client that they are sufficient and adequate.

Duty in relation to the health and safety file

Regulation 17 requires the client to provide all the health and safety information in his possession to the CDM co-ordinator for the purposes of the health and safety file. The information is more comprehensive than the information required by regulation 10 which relates to pre-construction information. Thus, the client should review the information that has been provided to the CDM co-ordinator for the purposes of regulation 10 prior to the construction phase to satisfy him- or herself that no other health and safety information has become available that is relevant for the health and safety file.

The CDM co-ordinator has the duty to prepare the health and safety file but it is the client's duty to take custody of the health and safety file and make it available as required by regulations 17(3) and (4). A full analysis of the requirements to comply with the preparation and custody of the health and safety file is discussed in Chapter 14.

Basic checklist of considerations for a client

		Regulation
Do the Regulations apply?		
1.	Does the content of the proposed project come within any of the definitions for construction work? Note that if they do, the Regulations apply.	
Identity of the client		
2.	Are there other persons who may fulfil the role of client?	
3.	Who will be the client for the purposes of the Regulations?	

		Regulation
4.	Do you wish to make the written election to be identified and fulfil the role of client?	8
Non-notifiable projects		
	Pre-construction phase	
5.	Is the structure to be used as a workplace?	10(2)(b)
6.	Have you gathered the pre-construction information in your possession after making appropriate searches and enquiries including:	
	• any information about or affecting the site or the construction?	10(2)(a)
	• any information concerning the proposed use of the structure as a workplace?	10(2)(b)
	• the time allocated for the pre-construction before commencement of construction work?	10(2)(c)
	• an existing health and safety file in respect of the structure?	10(2)(d)
7.	Do you need to appoint any designers?	9(1)
8.	Have you assessed the competence of the designers you wish to appoint?	4(1)
9.	Have you identified the appointments for contractors?	9(1)
10.	Have you assessed the competence of such contractors?	4(1)
11.	Have you ensured that all the designers and contractors who have been, or may be appointed by you, be provided with pre-construction information?	10(1)
12.	Have you made provision within the project documentation for the parties to co-operate with you and other persons?	5
13.	Have you provided within the project documentation the requirements and provisions for co-ordination with you and other parties?	6
14.	Are you intending to fulfil the role of any other duty holders?	

		Regulation
15.	If you are intending to fulfil the role of a duty holder, have you assessed your own competence?	4(1)(b)
16.	What reasonable steps have you taken to ensure that the arrangements made for designing a workplace have taken account of the provisions of the Workplace (Health, Safety and Welfare) Regulations?	9(1)(c)
	Construction phase	
17.	Are there any particular hazards or risks that should suggest the need for a 'quasi' construction phase plan?	9(1)(a) ACOP
18.	What reasonable steps have you taken to ensure that the arrangements for managing the construction work are suitable to ensure there is no risk to the health and safety of any persons?	9(1)(a)
19.	What reasonable steps have you taken to ensure that the welfare requirements of Schedule 2 are complied with for all persons carrying out construction work?	9(1)(b)
	Post-construction phase	
20.	Have you updated an existing health and safety file?	17(3)
Notifiable projects		
Pre-construction phase		
In addition to the above checklist for the pre-construction phase under non-notifiable projects, the additional points should be considered:		
21.	Have you assessed the competence of a CDM co-ordinator and made an appointment in writing?	14(1)
22.	Have you assessed the competence of a principal contractor and made an appointment in writing?	14(2)
23.	Have you provided the pre-construction information to the CDM co-ordinator?	15
24.	Have you ensured that the principal contractor has prepared a construction phase plan which complies with regulations 23(1)(a) and 23(2)?	16

		Regulation
25.	Have you ensured that the principal contractor has made provision for complying with the welfare requirements as set out in Schedule 2?	16(b)
Construction phase		
26.	What reasonable steps have you taken to ensure that the arrangements for managing the construction work are suitable to ensure there is no risk to the health and safety of any persons?	9(1)(a)
27.	What reasonable steps have you taken to ensure that the welfare requirements of Schedule 2 are complied with for all persons carrying out construction work?	9(1)(b)
Post-construction phase		
28.	Have you provided the CDM co-ordinator with all health and safety information in your possession relating to the project?	17(1)
29.	Have you ensured that the health and safety file is compiled such that each site or structure can be easily identified?	17(2)
30.	Is the health and safety file available for inspection by any person who may need access to it?	17(3)(a)
31.	Have you revised the health and safety file as often as may be appropriate to incorporate any relevant new information?	17(3)(b)
32.	If you have disposed of your entire interest in the structure, have you delivered the health and safety file to the acquirer together with information such that he is aware of the nature and purpose of the file?	17(4)

9 The designer

Who is a designer?

A designer is defined in regulation 2(1) as meaning:

> *any person (including a client, contractor or other person referred to in these Regulations) who in the course or furtherance of a business –*
>
> *(a) prepares or modifies a design; or*
>
> *(b) arranges for or instructs any person under his control to do so;*
>
> *relating to a structure or to a product or mechanical, electrical system intended for a particular structure, and a person is deemed to prepare a design where a design is prepared by a person under his control.*

Thus, any person can be a designer. In fact, many persons contribute to design unwittingly without recognising they are taking part in the design process.

What is design?

The definition of design in regulation 2(1) includes:

> *drawings, design details, specification and bills of quantities (including specification of articles or substances) relating to a structure and calculations prepared for the purpose of design.*

It is within most people's contemplation that design would include the preparation of drawings, calculations and specifications, but note, in particular, that the preparation of bills of quantities is included within the definition of design. The inclusion of bills of quantities hints at the importance of the allocation of financial resources which had been omitted from the 1994 Regulations and indeed which remains absent from the new Regulations.

Even in the most straightforward circumstances of preparing or modifying a design, a person may not be aware that their contribution will be part of the design process. In fact, because designs are rarely finalised until the point of installation or construction, all persons involved in the decision making up to the point of installation or construction are likely to be a designer. Therefore, in addition to the usual professions of architects, civil and structural engineers, building and quantity surveyors, landscape architects, principal contractor and contractors, the foreman or site engineer who modifies the design to achieve buildability by overcoming some local problems will be a designer.

There is no distinction between permanent works or the temporary works required for construction which will either be taken down or left within the fabric of the structure. The design may also include a product or mechanical or electrical system which is a component of a structure. Such components can be designed as part of a manufacturing process to be made as part of a manufacturing process.

The design does not have to be a neat set of calculations or instructions as the ACOP recognises that design can be communicated orally.

The definition of designer in regulation 2(1) is very widely drawn and the ACOP assists in categorising designers as set out below:

- Architects, civil and structural engineers, building surveyors, landscape architects, technical consultants, manufacturers and design practices associated with the professional mentioned and contributing to, or having overall responsibility for, any part of the design.

- Any person who specifies or alters a design, or specifies the use of a particular method of work or material.

- Building service designers, engineering practices or other persons designing plant which forms part of the permanent structure.

- Persons purchasing materials where the choice of material has been left to the discretion of the purchaser.

- Contractors carrying out design work as part of their contribution to a project.

- Engineers designing temporary works.

- Interior designers, including shop fitters and associated trades who develop the design.

- Heritage organisations who specify materials and method statements.

- Persons involved in making decisions as to the refurbishment or alteration of buildings and structures.

The ACOP clarifies that manufacturers supplying standardised products are not considered to be designers for the purposes of the Regulations, although inevitably they will have duties under sale and purchase legislation. However, the person who selects a product as a designer must take account of health and safety issues arising from its use in construction. If a product is made or altered for a particular project, the person who prepares the specification for the purpose-made product or alteration is a designer and so will also be the manufacturer who developed the detail design.

The definition of design will include, for the purposes of preparing or modifying a design, the following:

(i)   selection of design philosophy and principles;

(ii) specification of materials;

(iii) dimensions;

(iv) adoption of loading criteria;

(v)   method statements.

The definition of design and the role of designers are not changed in any respect according to the purpose of the design. A designer is subject to the same duties whether the design is for permanent structures, temporary works, modifications or maintenance works.

Arranging for or instructing persons under the designer's control

In sub-paragraph (b) of the definition for a designer, the concept of a person referred to as a designer actually preparing or modifying a design is not a difficult concept. It should not be difficult to recognise a person as a designer if they produce drawings, design details, specifications and bills of quantities. However, the term 'designer' as applied to other persons who merely arrange for others under their control to prepare a design is of limitless scope. The Regulations do not specify the type or nature of the required control. It cannot be assumed that the control referred to is limited to the ability or right of the arranging designer to control directly the work of preparing a design by others. Control can arise from various other situations, for example the control conferred by a majority shareholding in a subsidiary company; or contractual control (to the extent of fixing remuneration, programme and termination is not necessarily control over the design input or process); or a partner in a partnership.

The 1994 Regulations had set out in some detail a legal test for the arranging for or instructing any person under the designer's control. The Regulations no longer have a test for arranging or instructing. The normal meaning and understanding of the terms 'arranging for or instructing' are generally well understood and will be interpreted as such on the facts of each case as and when the situation arises. In respect of a design prepared or modified outside of Great Britain the test is of 'commission' and not arranging or instructing – see below.

Selecting a designer

Regulation 4(1)(a) provides:

> *No person on whom these Regulations place a duty shall appoint or engage ... a designer ... unless he has taken reasonable steps to ensure that the person to be appointed or engaged is competent.*

No single person on a project is charged with the responsibility to appoint a designer. The fact that design touches most functions and disciplines involved in the construction work would mean that it would be inappropriate to have any one named person with that responsibility. On large and complex projects there may be many different organisations appointing designers. There is only one factor to be taken into account by any person appointing a designer and that is the designer's competence.

Competence

In addition to the duty on a person who will be appointing or engaging a designer to ensure that the designer is competent, pursuant to regulation 4(1)(a), the designer is under a duty not to accept such an appointment or engagement unless he is competent as provided by regulation 4(i)(b) as follows:

> *No person on whom these Regulations place a duty shall accept such an appointment or engagement unless he is competent.*

To ensure that there is no likelihood there can be any defence that a designer was not subject to the duty to be competent, regulation 4(1)(c) provides:

> *No person on whom these Regulations place a duty shall arrange for or instruct a worker to carry out or manage design or construction work unless the worker is –*
>
> *(i) competent or;*
>
> *(ii) under the supervision of a competent person.*

In assessing the competence of designers, the person who intends to make the appointment or engagement should follow the two-stage assessment as suggested by the ACOP. Most importantly, before a designer can appreciate the risks involved in the project and how the design might serve to mitigate or eliminate those risks, the designer must have the skills and experience to identify the hazards.

The enquiries which a person should make of the designer to assess competence would include the criteria referred to as Stage 1 in the ACOP and set out in Appendix 4.

In making enquiries as to the criteria for hazard elimination and risk control, the Stage 2 criteria for part of assessing competence the appointer should consider in particular the following issues.

1. What part of the design has been, or is intended by the designer to be, carried out by sub-contractors together with, where appropriate, relevant information as to the competence of such sub-contractors?

2. What technical facilities are available to support the designer, particularly in the circumstances of the project? These may include:

    (a) computer-aided design facilities;

    (b) technical library;

(c) laboratory services;

(d) access to research information;

(e) membership of professional or trade associations;

(f) document/drawing storage.

3.   How will the designer ensure compliance with regulation 13(4)?

4.   What method will the designer use to communicate design information to ensure compliance with regulation 13(6)?

The client as designer

A client is not prevented from fulfilling the role of designer subject to being able to demonstrate that he or she is satisfied that his or her design capability, in house or otherwise, is competent and able to comply with the duties imposed by the Regulations. In circumstances where there is an independent CDM co-ordinator, it would be prudent for a client to seek his or her views before confirming the appointment of the designer from within his or her own organisation.

In some circumstances a client may wish to impose particular design principles, standards or specifications on a designer to ensure consistency of design with other projects which the client is procuring, or has procured. While the liability for design in such situations becomes blurred, the client will be fulfilling the role of designer under the Regulations. A client who has 'control' over the designer has to accept that he will become subject to the requirements of regulation 11 to an extent which will depend on the precise level of contribution to the design. It is recommended that the definition of a designer is studied carefully.

The CDM co-ordinator as designer

The role of CDM co-ordinator is inescapably linked with the design process. Therefore, while the CDM co-ordinator is not prevented by the Regulations from fulfilling the role of designer, there is the risk that there might be a lack of independence and objectivity to review the design considerations and produce adequate information.

Many design organisations will offer services as the CDM co-ordinator and there may be particular advantages to appointing the same organisation as designer and CDM co-ordinator. However, where possible, to preserve the 'independence' and objectivity of the CDM co-ordinator it would be preferable to have the CDM co-ordinator either in a different office or location from the designer or working within a completely different subsidiary company within the design organisation's group holding.

A contractor as designer

The Regulations have been drafted to accommodate the diverse increasing methods of procurement now used in the construction industry including 'design and build'. Under the forms of contract for design and build the main contractor undertakes to design the permanent and temporary works in addition to the execution of the construction phase. The exact extent of liability as to design will be subject to the scope of the contract terms. The Regulations do not prohibit or discourage a design and build contractor from fulfilling both roles as this is now a commonplace arrangement.

A contractor without any in-house design capability, whether for temporary or permanent works, will almost certainly need to appoint a designer. The assessment of competence of designers applies equally to temporary and permanent works design procured by a contractor.

Duties of designers

General duties for all projects

Designers have a general duty under the Management of Health and Safety at Work Regulations 1999 to undertake risk assessments as they relate to the design of the structure. The ACOP indicates that a designer who has complied with regulation 11 of the Regulations will have achieved compliance with regulations 3(1), (2) and (6) of the Management Regulations.

There is no express duty upon a designer to notify the client as to the existence of the client's duties under the Regulations. However, a designer is prevented from starting any design work until he knows that the client is aware of his duties under the Regulations. This duty is created by regulation 11(1) as follows:

> *No designer shall commence work in relation to a project unless any client for the project is aware of his duties under these Regulations.*

Unless the designer already has evidence that the client is aware of his duties under the Regulations, the designer should enquire as to whether the client is aware of his duties under the Regulations. The obligation in regulation 11(1) is absolute and the designer has no defence if he starts design work without knowing that the client is aware of his duties. In making enquiries of the client the designer will have put the client on notice that he should be complying with the Regulations and can start design work without risk. It follows, of course, that if the CDM co-ordinator has been appointed, the designer can safely assume that the client is aware of his duties. Unless the designer's instructions affirm that the client is aware or there is an acknowledgement from the client himself that he is aware of the Regulations, the designer should not commence the design work.

The duties to which the designer is required to comply with in preparing or modifying a design are subject to the test of reasonably practicable as provided for by regulation 11(2) as follows:

> *The duties in paragraphs (3) and (4) shall be performed so far as is reasonably practicable, taking due account of other relevant design considerations.*

The need to take account of other design considerations is a particular challenge for designers to decide what expense and effort needs to be expended by the designer to be proportionate within the framework of reasonably practicable.

The difficulty for the designer lies in identifying and appreciating the consequences of designs undertaken by others and transient conditions during the construction phase. If timing of activities is not identified by the designer at the outset, a moderate risk during the construction phase may in fact become a serious risk due to the timing of say deep excavations or lifting of heavy plant into place. For this reason, the designer should keep the design process under review during the course of construction, taking particular note of the programming of activities and considering carefully the designs of others as they also develop.

The designer's general duty under the Regulations is to prepare or modify designs which do not present a health and safety risk during construction or maintenance to construction workers and the public at large, those workers in the future who will be engaged in cleaning and maintaining the structure and workers using a structure as a workplace. Regulation 11(3) creates the duty as follows:

> *Every designer shall in preparing or modifying a design which shall be used in construction work in Great Britain avoid foreseeable risks to the health and safety of any person –*

(a) carrying out constructions work;

(b) liable to be affected by such construction work;

(c) cleaning any window to transparent or translucent wall, ceiling or roof in or on a structure;

(d) maintaining the permanent fixtures and fittings of a structure; or

(e) using a structure designed as a workplace.

Clearly, the duty extends to the construction work involving activities, the appreciation of which the designer must be able to demonstrate as part of his competence. Unless a designer can understand the manner in which a structure will be constructed, the designer will not be competent.

The designer has to consider the effect of his design on construction workers, persons on adjacent sites and the public at large.

A risk assessment, carried out by a designer, should follow the normal pattern of the following steps:

1. Identification of hazards;

2. A qualitative or quantitative assessment (depending on the circumstances) of the risks associated with each hazard.

The detail and amount of time a designer will devote to a risk assessment for a particular structure will depend upon its scale and complexity, but the obligation to identify foreseeable risks is an onerous one. A foreseeable risk has to be associated with a foreseeable hazard. Therefore, unless the designer is able to identify all the foreseeable hazards, not all the foreseeable risks will be included in the assessment. Note that what is, or is not, foreseeable is not subject to any reasonableness criterion, which is to say that in the event of an unusual accident to a person on a project, the risk was foreseeable, however small. The lack of a reasonableness criterion places a very high burden of responsibility on designers. Although a designer is required to have identified all foreseeable risks, the only requirement on the design is to have 'adequate regard' for such risks. Therefore, it is not a requirement of a design to eliminate all foreseeable risks. Understanding the creation and existence of hazards and the appreciation of risk emphasises the need for competent designers who can draw on their own experience and published information to comply with regulation 11(3).

The designer may find it helpful in conducting a risk assessment exercise to break the structure down into three main sources of hazard. First, the designer should examine methods by which the structure might be built in the context of his design options. The need to avoid risks completely, or to tackle them at source, or by reducing or controlling their effects will impact on design decisions. Despite the need to assess the risks during the construction phase associated with the method of construction, the Regulations do not require designers to specify construction methods or to exercise a health and safety management function over contractors, as they carry out construction work. In certain projects, the method of construction may be inextricably linked to the design philosophy, in which case the contractor could be contractually obliged to comply with such method of construction set out in the contract documents.

Second, the designer will have responsibility for specifying substances or equipment for use during the construction work. The manufacturers of such substances and equipment are obliged, by virtue of HASWA 1974, to produce product literature identifying any risks in handling or use, which should be considered by the designer. Not all hazards can be avoided, such as alkaline burns arising from skin contact with wet concrete, but prevention is easily effected provided that the information is communicated to the relevant persons.

Finally, the designer also has an obligation to consider persons who will be engaged in construction work and cleaning work on the structure after the completion of the construction phase. Particular hazards will include, for example, falling from height during cleaning operations, electrocution or electric burns when repairing or maintaining electrical equipment, etc.

Since no design is ever likely to be free of risk, there will always be residual risks even after the final design solution has taken adequate regard of all foreseeable risks. The remaining risks can be eliminated or reduced by combating at source the risks to the health and safety of workers. For example, many accidents due to falls during construction and cleaning work could be reduced by the pre-planning and design of cast-in eyebolts for the attachment of harnesses. Identifying problems associated with the build-up of exhaust gases or other fumes which can be overcome by improving ventilation in confined spaces would be another example. The examples are almost endless and are beyond the scope of this book. However, the increasing awareness of the influence of design over health and safety management of projects will be improved by a growing body of published literature and statistics.

The avoidance of risk will, in many cases, be impossible. The risks which the designer has to avoid are those which are foreseeable. On the basis that hazards will arise during construction and be created by the physical attributes of the construction site and the construction processes, it is extremely difficult to identify what risks may not be foreseeable. The very fact that a risk is not foreseeable would mean that, after the most comprehensive brainstorming, the risk associated with a hazard had not been identified. Although the test of foreseeability was included at the very last stage in the draft of the Regulations, it only contributes the slimmest of defences to any designer seeking to avoid liability under the Regulations.

Once a risk has been identified as being foreseeable, that risk has to be avoided. In practice, the avoidance of risks will, in many cases, be impossible. Risks associated with hazards can only be minimised, and rarely eliminated, if ever. The challenge for the designer is to take account of all the foreseeable risks associated with a particular hazard and then apply the necessary design steps to avoid that risk, so far as is reasonably practicable having taken due account of regulation 11(2).

The duty on the designer to avoid foreseeable risks puts the designer in an influential position with respect to the outcome of a project. If the client or principal contractor or others required a design change which, in the opinion of the designer would be likely to create or increase a risk, the other duty holders involved in the project would proceed with such a change at considerable risk to their own liabilities.

The designer is confronted with the duty to avoid unforeseeable risk at the outset of a design, save for initial design. Other than initial design which might include competition bids, applications for grants or feasibility studies the designer has to be considering health and safety from the outset. Even where initial design work may be involved in feasibility studies the health and safety aspect should not be ignored if a true comparison of the alternatives is to be undertaken. In an obvious example of comparing the options for a tunnel project, which could be carried out by compressed air tunnelling or deep open cut, the final design assessment would not be comprehensive without an appreciation of the risks to health and safety involved in compressed air tunnelling with working at the bottom of deep excavations.

In taking all reasonable practicable steps to eliminate risk, the designer has to take into account the likelihood of harm, the potential severity of the harm, the number of people potentially exposed to the harm and the period of duration or the turn period when the hazard would exist. Factored against those considerations are the costs, both design costs and construction costs, which would be necessary to eliminate such risks.

Designers will say that such considerations are likely to limit creativity and innovation. In the case of *IBA* v. *EMI*[1] it has been established by the Courts that a designer has no defence merely because the project was innovative. The case arose from the collapse of an aerial mast 365 m high constructed on Emsley Moor in Yorkshire. It was one of the highest masts in the world. Although the House of Lords recognised that the design approach put it 'beyond the frontier of professional knowledge at that time' when balanced against the potential consequences of a collapse, a very high standard of care was required. One of the Law Lords said 'the law requires even pioneers to be prudent' thus the case has important lessons for engineers embarking upon innovative design especially from a health and safety perspective.

The designer has to consider at the outset the application of minimising risks during cleaning and maintenance. The remainder of regulations 11(3)(c) and (d) impose upon the designer the same duty to consider cleaning and maintenance. A designer should be able to demonstrate as part of the design process that consideration has been given to cleaning including options for cleaning and maintenance which either may become part of the design or a recommended procedure to the client once the structure is in use.

The Regulations recognise that some hazards can be eliminated but that this may not always be possible. Regulation 11(4) states:

> *In discharging the duty in paragraph (3) the designer shall –*
>
> *(a) eliminate hazards which may give rise to risks; and*
>
> *(b) reduce risks from any remaining hazards,*
>
> *and in so doing shall give collective measures priority over individual measures.*

The designer is required to give collective measures priority over individual measures. The ACOP does not provide any examples of collective measures and neither does the industry guidance. However, if there is an unavoidable hazard of working at height, the designer should consider more than just one means of eliminating or minimising the risk. The designer might, for example, consider the installation of a temporary platform and safety harnesses attached to designated points on the structure. Each additional measure when applied in combination with another will have the effect of reducing risk exponentially.

Workplace considerations

The designer's duties go beyond considering factors such as appearance, maintenance, buildability and costs, to include the positive obligation to consider how the structure will perform as a workplace. Regulation 11(5) provides:

> *In designing any structure for use as a workplace the designer shall take account of the provisions of the Workplace (Health, Safety and Welfare) Regulations 1992 which relate to the design of, and materials used in, the structure.*

Workplaces include any place where workers can be found working. Regulation 2(1) provides that the definition of 'workplace' means a workplace within the meaning of regulation 2(1) of the Workplace (Health, Safety and Welfare) Regulations 1992 as follows:

> *any premises or part of premises which are not domestic premises and are made available to any person as a place of work, and includes:*
>
> *(f) any place within the premises to which such person has access while at work; and*
>
> *(g) any room, lobby, corridor, staircase, road or other place used as a means of access to or egress from that place of work or where facilities are provided for use in connection with the place of work other than a public road;*
>
> *but shall not include the modification, an extension or a conversion of any of the above until such modification, extension or conversion is completed.*

The designer therefore has to take account of the risks related to the proposed occupational use of the structure. The definition of workplace includes access and egress and therefore the designer should take account of external works including traffic and pedestrian routes. The designer should also consider the health and safety of the workers during any activities of cleaning and maintaining the permanent fixtures and fittings.

In designing so that the risks to the users and occupants of a structure can be eliminated or minimised to acceptable levels, the designer should ideally consult with the client to ensure that he or she has adequate information about the use and occupancy of the building provided as part of the information to be provided pursuant to regulation 10.

Communication, co-operation and co-ordination

Any design has to be communicated if it is to be implemented. The Regulations also recognise that co-operation and co-ordination of all activities including design are vital to reducing risks. The designer's duty to communicate is set out in regulation 11(6) as follows:

> *The designer shall take all reasonable steps to provide with his design sufficient information about aspects of the design of the structure or its*

construction or maintenance as will adequately assist –

(a) clients;

(b) other designers; and

(c) contractors,

to comply with their duties under these Regulations.

The other duty holders involved in a project need to know that their own proposals for designs and execution of the construction work are not going to be adversely affected by a component of the design that does not obviously create other inadvertent risks. The designer has to provide the sufficient information such that those who need the information, and who may not be of the same professional discipline, can easily understand how the design may impact on their own contribution to the project. The information should include:

- design philosophy;

- specification of materials;

- dimensions;

- adoption of loading criteria;

- conditions of temporary instability;

- method statements.

The ACOP suggests that information should be brief, clear and precise and can be provided by means of notes on drawings, written information provided on the design and suggested construction sequences.

The sufficient design information should be communicated before other related designs are completed and certainly before construction work commences.

The duty of co-operation imposed upon all duty holders by virtue of regulation 5 which compels designers to co-operate with the client, other designers and contractors. It is only by this means that any incompatibilities or inconsistencies between designs can be identified and resolved before the commencement of construction work.

The distribution of the design assumptions at an early stage should be encouraged so that all designers are working from the same starting-point. The designer should encourage design meetings and be enthusiastic to attend design meetings as part of his duty of co-operation.

The designer should continue to review hazards and risks arising from the progress of the construction work as they affect the designer's own design. The designer's role to participate in the co-ordination envisaged by regulation 6 should not be overlooked or underestimated.

Additional duties for notifiable projects

If a project is notifiable the designer must not start design work, other than initial design work, connected with the project unless a CDM co-ordinator has been appointed. Regulation 18(1) states as follows:

> *Where a project is notifiable, no designer shall commence work (other than initial design work) in relation to the project unless a CDM co-ordinator has been appointed for the project.*

On projects where the designer is aware that the scope of work would make the project notifiable, the designer must satisfy him- or herself that the CDM co-ordinator has been appointed before starting design work. However, for many projects the scope of work may be such that such an assessment is less obvious. In those circumstances, the designer is in a position to make a judgement as to whether the project should be notifiable. In those circumstances, the designer would be advised to approach the client to discuss the likely scope of the project and the criteria for notification. While this is not an express regulatory duty, the prudent designer would be advised to make such an approach because the ambit of regulation 18(1) provides no defence to a designer in the event that the project was notifiable and no CDM co-ordinator had been appointed.

The designer will know that if there is a CDM co-ordinator the project is notifiable and pursuant to which the CDM co-ordinator has various duties. Many of the duties of the CDM co-ordinator are reciprocated with duties from the other duty holders. The designer's reciprocal duties are set out in regulation 18(2) as follows:

> *The designer shall take all reasonable steps to provide with his design sufficient information about aspects of the design of the structure or its construction or maintenance as will adequately assist the CDM co-ordinator to comply with his duties under these Regulations, including his duties in relation to the health and safety file.*

The designer's duty is to provide 'sufficient information'. Unless the CDM co-ordinator is of the same professional discipline, it is unlikely that the CDM co-ordinator will be in a position to assess whether the information provided by the designer is sufficient. For these reasons, the responsibility for determining the sufficiency of the information lies with the designer.

The designer has to take all reasonable steps to provide the information. This will include editing and presenting the design data so that the information provided can be easily understood. It also means that there will be no defence to a designer if having posted the information it did not reach its destination. When providing the information it should ideally be part of a dialogue resulting in a confirmation, preferably in written form, from the CDM co-ordinator that the information has been received and that the CDM co-ordinator's requirements have been satisfied by the designer.

The designer's information about aspects of the structure should include the design principles. The design principles are likely to have a bearing on the construction particularly with regard to requirements for temporary works design to overcome potential instability during construction. The issues that arise from maintenance of the structure and to the extent that the design has an influence over maintenance, this information should also be provided to the CDM co-ordinator. In this regard, the designer should take into account the potential loadings which may be applied to the structure during maintenance and thus provide notification as to safe working loads for various parts of the structure.

The duties of the designer to provide the information pursuant to regulation 18(2) are an extension of his duty under regulation 11(6).

Designs prepared or modified outside of Great Britain

In the global marketplace it is now not uncommon for the location where a design is prepared or modified to be on the other side of the world. The effect of the Regulations would be undermined if designs that have been prepared or modified overseas could be imported into Great Britain as the basis for construction work to be executed within England, Wales or Scotland without having to comply with the Regulations. For that reason, the Regulations impose a filter or safety net to ensure that such designs are compliant with regulation 11. The relevant provisions are set out in regulation 12 which states:

> *Where a design is prepared or modified outside of Great Britain for use in construction work to which these Regulations apply –*
>
> *(a) the person who commissions it, if he is established within Great Britain; or*
>
> *(b) if that person is not so established, any client for the project,*
>
> *shall ensure that regulation 11 is complied with.*

The Regulations place the responsibility upon the person who commissions the design if they are established within Great Britain. This raises two issues which need to be tested. In the 1994 Regulations, the responsibility fell upon a designer who arranged for a person under his control to prepare or modify a design. The new test is, 'who commissions the design?' Plus, the element of control that was present in the 1994 Regulations has been omitted. A commission can amount to merely requesting that a design is prepared or modified.

The second test relates to the extent to which a person commissioning the design is 'established within Great Britain'. One might have expected the test to be related to having a place of business in Great Britain but the concept of being established has a much wider connotation. This may include key personnel of the person commissioning the design having a place of residence in Great Britain. Alternatively, the person commissioning the design may have assets within Great Britain but unconnected with their business interests. In either case, it means that a person can be served with proceedings within Great Britain or assets can be seized subject to an order of the Court.

In the event that the person commissioning the design is not established within Great Britain, the responsibility to ensure that the design complies with regulation 11 falls upon the client.

Basic checklist of considerations for a designer

All projects	Regulation
1. Have you checked that the client is aware of his duties?	11(1)
2. Are you satisfied of your competence to undertake the design (refer to stages 1 and 2 of the ACOP criteria)?	4(1)(b)
3. Have you checked the competence of any sub-designers working for you?	4(1)(a)
4. Have you received the health and safety file (if any) and the information you need to complete the design?	10
5. Have you identified hazards and foreseeable risks to those involved in the construction and future use of the structure?	11(3)
6. Have you established how your design can be constructed without risk and without adverse affect on the health and safety of construction workers, cleaners, occupants and general members of the public?	11(4) and (5)

All projects	**Regulation**
7. Have you provided sufficient information about specific aspects of the design that will assist clients, other designers and contractors?	11(6)
8. Have you co-operated with others involved in a project and identified those who need to co-operate with you?	5
9. Have you co-ordinated with others to complete your design for the construction work?	6
10. For the design of a workplace, have you taken account of the Workplace (Health, Safety and Welfare) Regulations 1992 and other relevant health and safety legislation?	11(3)(e) and 11(5)
11. Have you provided sufficient information about any significant risks connected with the design to other duty holders?	11(6)
Notifiable projects	
12. Have you checked that a CDM co-ordinator has been appointed?	18(1)
13. Have you taken all reasonable steps to provide design information that will assist the CDM co-ordinator in his duties particularly with respect to the health and safety file?	18(2)
What designers do not need to do	
14. Designers are not required to take into account or provide information about unforeseeable hazards and risk.	
15. The design carried out by the designer has only to take account of the foreseeable intended use of the structure in accordance with the design brief.	
16. The designer is not required to specify construction methods, except where the design relies upon a particular construction or erection sequence.	
17. The designer does not have any health or safety management function over contractors or others unless contractually bound to do so.	
18. A designer does not have to implement design measures to take account of trivial risks.	

Note

1. *Independent Broadcasting Authority* v. *EMI Electronics Ltd and BICC Construction Ltd* (1980) 14 BLR 1 HL.

10 The CDM co-ordinator

Introduction

During the years spanned by the 1994 Regulations, the Executive had been increasingly frustrated at the paperwork created by those fulfilling the role of the planning supervisor which resulted in increased costs for clients but without a commensurate improvement in the health and safety record in the construction industry. The Commission has finally conceded that the role of the planning supervisor under the 1994 Regulations had not been as successful as hoped for in contributing to an improvement in health and safety management.

Therefore, the Regulations have revoked the role of planning supervisor and have reworked the role in its new guise as the CDM co-ordinator such that, in the Commission'.s opinion, the CDM co-ordinator will be seen as the client's 'friend'.

The principal reason for the Temporary or Mobile Construction Sites Directive creating the role of CDM co-ordinator was to have some person in the project team able to provide advice and assistance to clients. One of the many challenges that face the construction industry is the enormous diversity of clients. Clients work within every possible industry sector and range from the highly sophisticated to the barely educated.

The new role in the Regulations ensures that the CDM co-ordinator retains a proactive role throughout a project including the construction phase.

By dividing projects into those which are notifiable and those which are not, and CDM co-ordinators only being a requirement for notifiable projects, the Regulations have created two very distinct regimes. By insisting on a CDM co-ordinator for the larger projects, the Regulations are acknowledging that a larger project usually involves more parties, more complexity and more inherent risk. However, many accidents occur on small projects which would not be notifiable but, in the Health and Safety Commission's judgement, the duties on the designer and contractor should be sufficient to protect the welfare of workers.

Appointment of the CDM co-ordinator

Regulation 2(1) provides:

> *CDM co-ordinator' means the person appointed as a CDM co-ordinator under regulation 14(1).*

The role of the CDM co-ordinator is mandatory on all projects that are notifiable. Moreover, the CDM co-ordinator can only be a person appointed by the client by virtue of regulation 14(1) as follows:

> *Where a project is notifiable, the client shall appoint a person ('the CDM co-ordinator') to perform the duties specified in regulations 20 and 21, as soon as is practicable after initial design work or other preparation for construction work has begun.*

During the project the client has an obligation to keep the appointment of the CDM co-ordinator under review by virtue of regulation 14(3) which provides:

> *The client shall ensure that appointments under paragraphs (1) and (2) are changed or renewed as necessary to ensure that there is at all times until the end of the construction phase a CDM co-ordinator ...*

The question arises as to when the client has to make the appointment. Clearly, the client has to be aware, in the first place, that the project will be notifiable. If the client is unaware of the criteria for notification this would be a matter that should be brought to the client's attention by the designer or contractor under their respective duties.

Regulation 14(1) requires that the appointment should be as soon as is practicable after initial design work. Initial design work can span very long periods of time and include various scheme designs and the consideration of different options from economic and technical points of view. Although the construction phase may seem a long way off at the stage of considering outline schemes, it will often be the case that one scheme would be much better than another from a health and safety point of view. As an example, the initial design might include options for a bridge or tunnel for a river crossing at its earliest stage. Why should the CDM co-ordinator not be appointed at this stage when the need to identify the respective hazards and risks to safety is just as relevant as economic and technical matters (not forgetting that these are all linked in any event)? If the option selected was for a tunnel, then at initial design the options for tunnel construction are certainly a matter for the CDM co-ordinator. As an example, the relative risks of construction by submerged tube or boring inevitably involve very different risk assessments.

While it can be seen that the CDM co-ordinator has a valuable contribution to make at the initial design stage, regulation 14(1) provides that the CDM co-ordinator may be appointed when other preparation for construction work has begun. This would imply that much of the design would have been completed and the difference in the time of appointment between initial design and preparation for construction work could be considerable. However, for the smaller projects, which are still notifiable, the principal contractor may be appointed in the role of CDM co-ordinator thus recognising the practicality that the distinction between initial design work and preparation for site work may be blurred.

Once a person has been appointed to the role of CDM co-ordinator, it should remain filled until the end of the construction phase when it is the responsibility of the CDM co-ordinator to pass the health and safety file to the client as provided for by regulation 20(2)(f).

Clients would be well advised to ensure that adequate provision is made in the terms of engagement of the CDM co-ordinator for a smooth handover to a replacement person, particularly with regard to notice of termination or change and identification of documentation to be handed over.

Who can be the CDM co-ordinator?

There is no limitation in the Regulations as to the type of person who may fulfil the role of CDM co-ordinator. The CDM co-ordinator can be an individual, partnership, limited company, local authority or government department, subject always to demonstrating the necessary competence and allocation of resources to the client.

The CDM co-ordinator has an obligation to decline an appointment if it does not consider itself competent to undertake the role by virtue of regulation 4(1)(b).

There is no prohibition within the Regulations which prevents the role of CDM co-ordinator being combined with one of the other roles under the Regulations, the main combinations of which are set out below. Combining any role with the CDM co-ordinator is likely to lead to a loss of objectivity and prejudice the effectiveness of the CDM co-ordinator's role, from a health and safety management point of view. The practice of appointing an independent CDM co-ordinator under separate terms of engagement and fee arrangements has distinct advantages in promoting objectivity and the clear duty to do what is necessary to improve health and safety, rather than what might be contractually expedient. For many projects, however, the size of the project will not warrant the additional cost of a specialist CDM co-ordinator but persons accepting such combined roles should recognise that the increased responsibility and liability is the price to be paid for the additional fee.

The client as CDM co-ordinator

A client can appoint itself as the CDM co-ordinator at the relevant time in accordance with regulation 14(1). Indeed, the default position if no CDM co-ordinator has been appointed is the deeming provision in regulation 14(4) that the client will be treated as though it is the CDM co-ordinator. However, the risk of sacrificing the objectivity which an independent CDM co-ordinator would have in fulfilling the requirements under regulation 14 should not be underestimated. The lack of objectivity is compounded if the client is also fulfilling one or more of the other roles.

A client has to be satisfied as to the competence of any person to be appointed as the CDM co-ordinator, as required by regulation 4(1). There is a risk that, despite the express requirement in regulation 4 as to competence, the client will not address his mind to the objective assessment of his own competence. A client is particularly at risk if the CDM co-ordinator has not been appointed or during any period for

which no person is fulfilling the role of CDM co-ordinator, as regulation
14(4) provides:

> *The client shall –*
>
> *(a) be deemed for the purposes of these Regulations, save paragraphs
> (1) and (2) and regulations 18(1) and 19(1)(a) to have been
> appointed as the CDM co-ordinator or principal contractor, or
> both, for any period for which no person (including himself) has
> been so appointed; and*
>
> *(b) accordingly be subject to the duties imposed by regulations 20 and 21
> on a CDM co-ordinator or, as the case may be, the duties imposed by
> regulations 22 to 24 on a principal contractor, or both sets of duties.*

The principal contractor as CDM co-ordinator

There is no prohibition in the Regulations which prevents the principal
contractor from being appointed as the CDM co-ordinator.

The CDM co-ordinator has, by virtue of regulation 14(1), to be
appointed by the client as soon as is practicable after the initial design
or other preparation for construction work has begun. By comparison
a client has to appoint a principal contractor when the client knows
enough about the project to be able to select a suitable contractor to
fill the role. The appointment of the principal contractor as the CDM
co-ordinator would imply that the principal contractor was expected
to be involved in the design stage following initial design or alternatively
preparation for construction work was urgent and able to proceed
concurrently with design work.

If a client wishes to appoint the same person to the roles of CDM
co-ordinator and principal contractor, at the outset of the project, it
would suggest from a commercial point of view that the role of principal
contractor would not be subject to competitive tendering at a later date.
For this reason, it is likely that clients appointing a principal contractor
at the outset will be more likely to be design and build contractors. For
large and complex projects the client should be aware of the risk that the
principal contractor may lack the competence to undertake the design in
accordance with the Regulations from within its own resources. This
could create a conflict between the principal contractor role as CDM
co-ordinator and the 'self-regulation' of its own design activities. A
client should always keep the issue of competence under review, and
yet the client would be lacking the independent and objective advice of
the CDM co-ordinator in such circumstances.

In keeping the appointments of CDM co-ordinator and principal contractor under review, the client can take advantage of combining the roles in the same person at any stage in the pre-construction or construction phase depending on all the circumstances.

A review of the perceived advantages and disadvantages of combining the role of principal contractor and CDM co-ordinator is given in Chapter 8 on the appointment of a CDM co-ordinator by the client.

A designer as CDM co-ordinator

One of the pressures that brought about a revision to the 1994 Regulations was the increasing awareness of the importance of the role of the designer in improving the health and safety environment for construction workers and users of a structure. In view of the widespread acceptance of the importance of the designer in achieving improvements in health and safety it would seem to be a good idea to appoint a designer as the CDM co-ordinator. The designer would bring to the role a detailed knowledge of the design process which is where other organisations may be less effective.

A distinct advantage of appointing a designer as the CDM co-ordinator is the fact that a designer is likely to be involved at the inception of a project. This would facilitate a proper health and safety risk assessment to be applied to early schematic options. As noted earlier, the identification of hazards as between different options could, and perhaps should, be a significant factor in assessing the outcome of the preferred solutions at the initial design stage.

Competence

There is no profession with a monopoly on producing CDM co-ordinators from within its ranks. Persons who are likely to be suited best to the role of CDM co-ordinator will include those who have an understanding of the design process and construction methods. This indicates that civil and structural engineers, architects, builders and building surveyors are likely to have appropriate backgrounds to enable them to fulfil regulation 20.

Any person who has attained chartered status in one of the professions previously mentioned will have achieved high academic standards. However, it is unlikely to be sufficient for a CDM co-ordinator to rely on chartered status without evidence of a knowledge of health and safety issues and management. Such evidence can be obtained by individuals taking further qualifications, by examination and assessment in occupational health and safety. However, the Regulations do not make it necessary for a CDM co-ordinator acting as an individual to be of chartered status or have any specialist qualifications. For smaller straightforward projects, the academic and professional qualifications may be adequate at the level of technician engineer or engineering technician. Academic and professional qualifications are not, on their own, enough. In addition to the academic qualifications a CDM co-ordinator must be able to demonstrate an appropriate level of experience which may be gained during time spent in an individual's 'first' profession. Appendix 5 of the ACOP indicates that the required level of academic qualifications for fulfilling the duties of CDM co-ordinator on large and complex projects will be of a high standard backed up by several years of varied and relevant experience. The guidance for assessing the competence of a CDM co-ordinator is reproduced in Appendix 4.

When assessing the competence of a partnership or corporate organisation the client should consider the resources that can be called upon to undertake the day-to-day management of the CDM co-ordinator's general duties. Inevitably, the individuals will be at the centre of any enquiries. A company that has no skills or experience in health and safety management at a level sufficient to demonstrate competence can acquire the necessary competence by employing or appointing an individual who is competent within the meaning of the Regulations.

A client seeking to appoint a prospective CDM co-ordinator is required, by virtue of regulation 4(1), to be satisfied as to the competence of the person to be appointed. Reasonable enquiries and advice are no longer an express requirement but would be well advised so that the client will be able to assert that it did not rely entirely upon the person's own declarations of competence. It is suggested that advice should be sought from independent professional sources. These may include professional institutions, learned societies and other persons who offer their services as CDM co-ordinators. A suggested checklist of the enquiries based on the ACOP which a prudent client should make of a person prior to their appointment as a CDM co-ordinator might include:

1. What academic qualifications does the CDM co-ordinator possess with particular regard to occupational health and safety?

2. What is the CDM co-ordinator's knowledge of construction practice, particularly in relation to the nature of the project? Is it, for instance, building, earthworks, demolition, etc.?

3. What is the CDM co-ordinator's familiarity and knowledge of the design function?

4. What is the CDM co-ordinator's knowledge of health and safety issues?

5. Is the CDM co-ordinator able to demonstrate achievement in management, particularly in the role of CDM co-ordination and liaison? Has he the ability to work with and CDM co-ordinate the activities of different designers and be a bridge between the design function and construction work on site?

6. Does the CDM co-ordinator have relevant experience of preparing health and safety files?

7. Does the CDM co-ordinator understand and appreciate the contractual obligations of the other parties to the project?

Additional enquiries that the client might consider as relevant to the overall health and safety management of a project could include the following:

1. What number of individuals and their respective qualifications and experience are allocated to the project, both internally and from other sources, to perform the various functions in relation to the project?

2. What management system will be used to monitor the correct allocation of people and other resources in the way agreed at the time these matters were finalised?

3. What management system will be used to co-ordinate the activities of the designers and collect information for the preparation of the construction phase plan?

4. What time will be allowed to key personnel to carry out the different duties of CDM co-ordinator?

5. What technical facilities are available to aid the staff of the CDM co-ordinator in carrying out their duties including:

    (a) office location and accommodation;

    (b) information technology;

    (c) technical library.

6. Can the CDM co-ordinator demonstrate the implementation and practice of quality assurance principles? Registration with one of

the reputable accreditation organisations would be a quick and efficient means of providing the confidence that the CDM co-ordinator has a management system incorporating items 2 and 3 above, although such registration is not absolutely necessary.

Duties of the CDM co-ordinator

Notification

Contracts of appointment between clients and CDM co-ordinators will often extend the duties and obligations of the CDM co-ordinator beyond those required by the Regulations. The CDM co-ordinator should be alert to, and avoid if necessary, the additional contractual duties and obligations. What follows is the extent of the statutory duties only.

The CDM co-ordinator is only appointed for projects which are notifiable and a project is notifiable if the construction phase is likely to involve more than 30 days or 500 person days of construction work, as provided for by regulation 2(3).

Once appointed, the CDM co-ordinator will know that the project is notifiable and does not have to consider if the project is not notifiable.

The first task of the CDM co-ordinator is to deal with the requirements of notification pursuant to regulation 21(1) as follows:

> *The CDM co-ordinator shall as soon as is practicable after his appointment ensure that notice is given to the Executive containing such of the particulars specified in Schedule 1 as are available.*

The CDM co-ordinator will require a sound grasp of the project before sending a notice to the Executive. The information that the CDM co-ordinator requires is set out in Schedule 1 to the Regulations as follows:

1. Date of forwarding.

2. Exact address of the construction site.

3. The name of the local authority where the site is located.

4. A brief description of the project and the construction work which it includes.

5. Contact details of the client (name, address, telephone number and any e-mail address).

6. Contact details of the CDM co-ordinator (name, address, telephone number and any e-mail address).

7. Contact details of the principal contractor (name, address, telephone number and any e-mail address).

8. Date planned for the start of the construction phase.

9. The time allowed by the client to the principal contractor referred to in regulation 15(b) for planning and preparation for construction work.

10. Planned duration of the construction phase.

11. Estimated maximum number of people at work on the construction site.

12. Planned number of contractors on the construction site.

13. Name and address of any contractor already appointed.

14. Name and address of any designer already engaged.

15. A declaration signed by or on behalf of the client that he is aware of his duties under these Regulations.

The requirements of regulation 21(1) acknowledge that not all the particulars listed in Schedule 1 may be available at the time of the notice. In particular, the selection of the principal contractor and other contractors may not be known if the project is still at design stage. Regulation 21(2) provides specifically for the situation as follows:

> *Where any particulars specified in Schedule 1 have not been notified under paragraph (1) because a principal contractor has not yet been appointed, notice of such particulars shall be given to the Executive as soon as is practicable after the appointment of the principal contractor, and in any event before the start of the construction work.*

In reality, the CDM co-ordinator is likely to have a lot less information than just the identity of the principal contractor when he first makes a notification, unless the start of construction work is imminent. The CDM co-ordinator should keep the notice under review and ensure the Executive is kept informed by updating the notice with respect to appointments or any changes in the duty holders. It should be noted that the information for designers and contractors includes those who are employed by parties other than the client.

To reflect the recent development of reliance on e-mails in business, regulation 21(3) permits notification by e-mail as follows:

Any notice under paragraph (1) or (2) shall be signed by or on behalf of the client or, if sent by electronic means, shall otherwise show that he has approved it.

While the notice may have been prepared by the CDM co-ordinator, the client is required to have approved it either by a signature or by some other means. Thus, if the notice, as received by the Executive, is not signed or otherwise approved by the client it will not be a valid notice. By this means, the client is forced to address some of the fundamental issues with respect to resources and overall health and safety management at the outset of the CDM co-ordinator's involvement.

There is no recommended format for the notification. However, a form, F10 (REV), can be used and is available from the Executive or can be completed online at https://www.hse.gov.uk/forms/notification/f10.pdf

In respect of construction work for rail projects where the Office of Rail Regulations is the enforcing authority, the CDM co-ordinator should also give notice to the Office of Rail Regulation as required by regulation 21(4) as follows:

Insofar as the project includes construction work of a description for which the Office of Rail Regulation is made the enforcing authority by regulation 3(1) of the Health and Safety (Enforcing Authority for Railways and other Guided Transport Systems) Regulations 2006, paragraphs (1) and (2) shall have effect as if any reference to the Executive, were a reference to the Office of Rail Regulation.

The CDM co-ordinator should be alert to the fact that a project may include work which comes within the jurisdiction of the Office of Rail Regulation and the Executive, in which case notice should be given to both enforcing authorities.

A record of the notice should be retained as proof at any later stage that the CDM co-ordinator had complied with regulation 21.

Advice and assistance

Regulation 20(1)(a) provides that:

The CDM co-ordinator shall give suitable and sufficient advice and assistance to the client on undertaking the measures he needs to take to comply with these Regulations during the project (including, in particular, assisting the client in complying with regulations 9 and 16).

Paragraph (1)(a) is the 'best friend' obligation of the CDM co-ordinator to advise and assist the client. Such advice and assistance should be on a basis of full disclosure and in good faith for the sole objective of helping the client. The client's duties in relation to arrangements for managing projects (regulation 9) and the start of the construction phase (regulation 16) are fundamental to the efficacy of the Regulations. Many of those organisations who will be clients will not have the necessary skills and knowledge within the construction industry to discharge those duties without the advice and assistance of the CDM co-ordinator. All the other duties under regulations 20 and 21 are effectively subordinate to this paragraph (1)(a) and the person fulfilling the role of CDM co-ordinator should not underestimate the extent of that duty.

The clients who do not have the knowledge and expertise to identify suitable organisations to undertake the duties of designers and contractors should receive advice and assistance from the CDM co-ordinator. In particular, the client may need advice and assistance in assessing competence and ensuring that the duty holders will be adequately resourced for the work for which they are responsible. The client can also expect to receive advice and assistance on the overall procurement strategy to ensure that the arrangements for managing the project are such that construction work can be carried out, so far as is reasonably practicable, without risk to the health and safety of any persons.

The client has a duty to ensure that the arrangements for managing projects are maintained and reviewed throughout the project (regulation 9) and, therefore, the CDM co-ordinator should keep such matters under review and bring any failings or shortcomings to the attention of the client during the construction project.

Collection of pre-construction information

The CDM co-ordinator's role in being the collection point for information is set out in regulation 20(2)(a) and (b) as follows:

> *Without prejudice to paragraph (1) the CDM co-ordinator shall –*
>
> *(a) take all reasonable steps to identify and collect the pre-construction information;*
>
> *(b) promptly provide in a convenient form to;*
>
> > *(i)   every person designing the structure, and*
> >
> > *(ii) every contractor who has been or may be appointed by the client (including the principal contractor),*
>
> *such of the pre-construction information in his possession as is relevant to each.*

The definition of 'pre-construction information' means the information described in regulation 15 where the project is notifiable.

The CDM co-ordinator's duty to take all reasonable steps to identify and collect the pre-construction information relies to a large extent on the CDM co-ordinator's knowledge and experience as to the type and nature of information that is required and is likely to be in existence. The CDM co-ordinator should be alert to identifying at an early stage omissions and gaps in the pre-construction information. Although the client has a duty to provide information pursuant to regulation 15, the CDM co-ordinator has a wider duty to identify and collect the pre-construction information from other sources.

Appendix 2 of the ACOP lists the information which should be included for the purposes of persons bidding for or planning work, and for the development of the construction phase plan which is reproduced in Appendix 2.

Once the information is in the possession of the CDM co-ordinator, he or she is required to communicate the information in a convenient form. The means by which the information is presented is likely to have a significant effect on the costs incurred by the receiving parties. The CDM co-ordinator can have an important influence at this stage of a project in mitigating wasted costs in sorting unstructured and uncollated information.

Co-ordination and co-operation

Co-operation and co-ordination is a general principle that underpins the philosophy of the regulations and the CDM co-ordinator has an important duty in facilitating such co-operation and co-ordination. The CDM co-ordinator's specific duties in this respect are set out in regulations 20(1)(b) and 20(2)(d).

In respect of the pre-construction phase, regulation 20(1)(b) provides that:

> *The CDM co-ordinator shall ensure that suitable arrangements are made and implemented for the co-ordination of health and safety measures during planning and preparation for the construction phase, including facilitating –*
>
> *(i) co-operation and co-ordination between persons concerned in the project in pursuance of regulations 5 and 6, and*
>
> *(ii) the application of the general principles of prevention in pursuance of regulation 7.*

The CDM co-ordinator should establish a mechanism or network for encouraging and promoting communication between the various parties to any project. This is a task which should be implemented as soon as possible because the planning and preparation during the pre-construction phase will have a profound effect on the management of the construction phase.

The challenge in putting in place a framework for communication is the need for monitoring communications so that the content and quality of communication is appropriate. This is particularly needed with respect to the application of the general principles of prevention.

The means of communication will reflect the preferences of the parties to the project and will be influenced by the proximity of the parties, number of parties and complexity of the issues.

The CDM co-ordinator should adopt an even-handed approach between the parties and gain their trust so that the flow of information is open, unrestricted and transparent.

The CDM co-ordinator can set an example for the parties to the project by disseminating the information which he will have received from the client and stimulating consideration of that information by making appropriate enquiries as to how the general principles of prevention are being incorporated and addressed.

In respect of the construction phase, regulation 20(2)(d) provides:

> *Without prejudice to paragraph (1) the CDM co-ordinator shall take all reasonable steps to ensure co-operation between designers and the principal contractor during the construction phase in relation to any design or change to a design.*

The CDM co-ordinator's understanding of the design process and how contributions to design require co-ordination to produce a final overall design is vital. To discharge the duty of regulation 20(2)(d) the CDM co-ordinator has to maintain communication with the designers and the principal contractor during the construction phase. Design changes will often occur during the construction phase to overcome unforeseen difficulties even when there is no outstanding design left to be completed after commencement of the construction phase. Thus, the CDM co-ordinator cannot ever assume that changes in design will not be necessary during the construction phase.

The designers will include persons appointed by the principal contractor and others appointed by those persons. The first task, therefore, for the CDM co-ordinator, is to recognise once the construction phase has started who is contributing to the design network. The principal contractor will often be tempted to modify designs to avoid delay and costs during the construction phase. The CDM co-ordinator should make site visits, at the very least, and check that the construction is proceeding in accordance with the agreed design. The CDM co-ordinator should also make regular enquiries of the principal contractor, usually at progress meetings, to ensure that there have been no design changes or that none are envisaged for the future. As design changes are identified, or planning the construction requires design changes, the CDM co-ordinator should ensure that these are communicated to all those other designers whose own designs may be affected.

The CDM co-ordinator's role in this regard is probably the most challenging duty. The CDM co-ordinator is unlikely to have the same level of expertise and design for all the elements of the project which should mean that the CDM co-ordinator is seeking the advice of the other designers as to the impact of any changes.

On large projects where the design was not completed at the start of the construction phase, the need to have continuing liaison over the developing design cannot be avoided. The CDM co-ordinator should ensure that the construction phase plan provides for the co-operation of the principal contractor and other contractors in liaising to obtain co-operation in finalising the design.

The CDM co-ordinator's contribution to and influence over design should not be understated. The following section on the CDM co-ordinator's duty with respect to designers in regulation 20(2)(c) emphasises the point.

The CDM co-ordinator and designers

Regulation 20(2)(c) emphasises the role which the CDM co-ordinator has in the overall design process as follows:

> *Without prejudice to paragraph (1) the CDM co-ordinator shall take all reasonable steps to ensure that designers comply with their duties under regulations 11 and 18(2).*

How is the CDM co-ordinator to know that a designer has not complied with his or her duties under regulations 11 and 18(2)? In particular, how can the CDM co-ordinator ensure that designers have identified all the hazards? The answers are that the CDM co-ordinator must have some knowledge and competence in the relevant design discipline. Of course, the CDM co-ordinator will have provided to the designers all the information they require so that the duties under regulations 11 and 18(2) can be discharged. There is no requirement that the CDM co-ordinator is obliged to check the design prepared by the designer but the duty to ensure compliance by the designer must involve some degree of checking. It is not possible to check the designer's compliance with regulations 11 and 18(2) without having a level of skill and expertise in that design discipline.

From a practical point of view, the CDM co-ordinator who has the necessary skill and expertise in the discipline will earn the respect of the designers and thus obtain their co-operation with considerably more ease than if the CDM co-ordinator has no or little background in design.

Liaison with the principal contractor

Although the CDM co-ordinator is required to ensure that suitable arrangements are made for the construction phase, he or she does not have to be involved in the preparation of the construction phase plan. However, the CDM co-ordinator cannot abdicate responsibility for the principal contractor's duty to prepare the construction phase plan as recognised by regulation 20(1)(c) as follows:

> *The CDM co-ordinator shall liaise with the principal contractor regarding –*
>
> *(i) the contents of the health and safety file,*
>
> *(ii) the information which the principal contractor needs to prepare the construction phase plan, and*
>
> *(iii) any design development which may affect planning and management of the construction work.*

The reference to the contents of the health and safety file will include any health and safety file that has been provided by the client during the pre-construction phase and the information provided at the end of the construction phase which the CDM co-ordinator needs so that he can prepare the health and safety file. Thus, (i) is the CDM co-ordinator's responsibility, whereas (ii) is the principal contractor's responsibility in respect of preparing the construction phase plan. The close liaison which is required between the CDM co-ordinator and principal contractor will ensure that the CDM co-ordinator is able to advise the client appropriately when the client is required to discharge his duty under regulation 16.

The principal contractor may not have the close relationship with designers who have been working upon the project during the pre-construction phase and the CDM co-ordinator is the vital link to ensure that design development does not become dislocated from planning and management of the construction work. Design development is an iterative process and the CDM co-ordinator is responsible for recognising developments in design and thereafter ensuring that they are addressed by the relevant parties.

The health and safety file

The CDM co-ordinator's duty includes the creation or updating of the health and safety file. The definition of the health and safety file and the duties are created by regulations 20(2)(e) and (f) as follows:

> *Without prejudice to paragraph (1) the CDM co-ordinator shall –*
>
> *(e) prepare, where none exists, and otherwise review and update a record ('the health and safety file') containing information relating to the project which is likely to be needed during any subsequent construction work to ensure the health and safety of any person, including the information provided in pursuance of regulations 17(1), 18(2), 22(1)(j);*
>
> *(f) at the end of the construction phase, pass the health and safety file to the client.*

The importance of the health and safety file is such that Chapter 14 deals with this topic in detail. However, for the CDM co-ordinator, the importance of the health and safety file can have far-reaching consequences. If the CDM co-ordinator is negligent in creating or updating a health and safety file, either by omitting critical information or incorporating inaccuracies, this is likely to impact on those who later come to rely upon the health and safety file. The persons who rely upon the health and safety file in a later project and in so doing suffer harm, or are the cause of some other person suffering harm, may have a cause of action against the CDM co-ordinator who was responsible for the health and safety file.

Basic checklist of considerations for a CDM co-ordinator

	Notification and information	**Regulation**
1.	Have you notification in writing from the client of your appointment or agreed written terms of engagement?	14(1) and (5)
2.	Have you notified the Executive of the project?	21
3.	Do you have an agreed line of communication to the client?	20(1)(a)
4.	Have you taken all reasonable steps to identify and collect pre-construction material?	20(2)(a)
	Competence and sufficient time and other resources	
5.	Have you considered your competence including familiarity and experience of the design process and resources before accepting the appointment?	4(1)(b)
6.	Have you demonstrated competence to the client?	4(1)(a)
7.	Have you allocated sufficient time and other resources?	9(1)
8.	Are you prepared and able to provide advice and assistance to the client on competence with respect to other members of the project team?	20(1)(a)
9.	Are you prepared and able to provide advice to the client on the allocation of sufficient time and other resources by the contractors in the project team?	20(1)(a)
	Health and safety of persons	
10.	Have you taken steps to ensure that designers comply with their duties with regard to regulations 11 and 18(2)?	20(1)(c)
11.	Have you ensured that suitable arrangements are made and implemented for the co-ordination of health and safety measures during planning and preparation for the construction phase?	20(1)(b)
12.	Have you taken steps to ensure co-operation between designers?	20(2)(d)

		Regulation
Construction phase		
13.	Have you ensured that the information specified in regulations 10 and 15 relevant to each person has been circulated promptly?	20(2)(b)
14.	Have you taken all reasonable steps to ensure co-operation between designers and the principal contractor during the construction phase in relation to any design or change to a design?	20(2)(d)
15.	Have you a line of communication to the principal contractor for the purposes of contributing to the development of the construction phase plan, the contents of the health and safety file and any design development which may affect planning and management of the construction work?	20(1)(c)
Health and safety file		
16.	Have you ensured that a health and safety file is prepared in respect of each structure?	20(2)(e)
17.	Have you kept the health and safety file under review up to completion of the construction phase?	20(2)(e)
18.	Have you handed the health and safety file over to the client on completion of the construction phase?	20(2)(f)

11 The principal contractor

Introduction

Accidents do not happen in the client's boardroom or in the design office
– when the planning and design are completed it is the principal contractor who is responsible for the safe execution of the construction works.

The roll call of contracting organisations who have been prosecuted
successfully under the 1994 Regulations reveals that breach of the
Regulations is not confined to the small, badly managed and 'here
today gone tomorrow' contractors. The fact that some of the contractors
who have been prosecuted have national and international reputations
suggests that close attention to the principal contractor's duties and
obligations is still necessary.

Definition

The principal contractor is defined in regulation 2(1) as meaning:

> *the person appointed as the principal contractor under regulation 14(2).*

The role of principal contractor is mandatory on all projects which are notifiable under the Regulations. Moreover, the principal contractor can only be appointed by the client by virtue of regulation 14(2) as follows:

> *After appointing ... a suitable person such appointment.*

The role and duties of a contractor are discussed in Chapter 12. However, for the purposes of discussing the role of the principal contractor, it should be noted that a contractor carries out or manages construction work. Thus, a client is able to appoint contractors who do not carry out any construction work but merely provide a management service. This extension of the traditional view of contractors takes account of the trend for management contracts and the use of project management organisations.

Selecting a principal contractor

A client who is required to appoint a principal contractor has to consider his competence in the same way as a client has before appointing a CDM co-ordinator, contractor or designer.

The client is not required to appoint a main contractor or management contractor as the principal contractor. In other words, the role of the principal contractor in executing the construction work is a relevant factor but not necessarily the only or determining factor.

The client's duty to ensure that the principal contractor is competent is matched by the principal contractor's duty not to accept such an appointment if he is not competent. This duty of implied self-assessment is stated in regulation 4(1)(b) as follows:

> *No person whom these Regulations place a duty [the principal contractor] shall accept such an appointment or engagement unless he is competent.*

The ACOP will be the primary source of information for a client in assessing competence of a principal contractor. The ACOP recognises that the client has to make a judgement as to whether the principal contractor has the competence to carry out the work safely. As if to mitigate the rigour of regulation 4, which requires the client to take reasonable steps to ensure that the principal contractor is competent, as opposed to being reasonably satisfied, the ACOP comforts clients that, if their judgement is reasonable, they will not be criticised. This assumes that they have taken the reasonable steps and taken into account the evidence that has been asked for and provided. However, with respect to the principal contractor's own assessment of his competence there is no such mitigation or comfort. In complying with regulation 4(1)(b), a principal contractor has to assess that he or she is either competent or not competent. Since the principal contractor's judgement is with respect to matters within his or her own knowledge and experience, that is not unreasonable. Nonetheless, the principal contractor should not underestimate this express requirement confirming his or her competence to undertake the particular project. The ACOP has provided a framework for the benefit of clients for the Stage 1 assessment of competence for contractors based on 12 criteria. None of the criteria are specifically for principal contractors. The Stage 2 assessment, which includes work experience, will highlight the previous experience in the role of principal contractor although clients should recognise that every contractor, aspiring to be a principal contractor and who has not yet been appointed as a principal contractor, needs to obtain its first appointment.

The principal contractor must provide evidence to the client in the honest belief that the evidence supports the principal contractor's own assessment. At the Stage 1 assessment, the principal contractor does not have to demonstrate experience, but by putting in place written policies and procedures supported by the evidence indicated in the ACOP, a client should be able to confirm the attainment of competence if the principal contractor's responses 'tick all the boxes'. Of course, merely having written policies and procedures is not evidence of competence in practice which is why the Stage 2 assessment is critical.

The principal contractor must be prepared to submit evidence of compliance with the policies and procedures by reference to projects of comparable size, content and complexity. The evidence can include independent certification and verification or the submission of specific documents which provide a review of internal records, updates, reviews and meetings.

The principal contractor should recognise, in addition to the self-assessment of its own competence, that it has the further responsibility of assessing the competence of its own workforce to carry out or manage construction work and the competence of its sub-contractors. From the client's point of view, there will inevitably be a period of time leading up to the appointment or engagement of the principal contractor. The client may have pursued a competitive tendering selection procedure or may have selected a principal contractor for a negotiated contract. The client has to consider carefully when the assessment of competence is concluded. Should it be before putting the principal contractor to the cost and expense of preparing and submitting a tender or even requiring the principal contractor to develop the construction phase plan to a specified level of detail before making the engagement?

The principal contractor will incur costs merely in providing the evidence to the client of his own competence and clients should balance those commercial costs against the convenience of the client's own programme for awarding a contract.

The enquires that should be made by a client, whether by the pre- or post-tender stage, certainly before the award of the contract, are listed below. The steps are based on the ACOP core criteria but go further in identifying those criteria that are suitable for pre-qualification prior to tendering.

Assessment of competence

Stage 1: The pre-qualification prior to tendering

1. *The existence of the principal contractor's health and safety policy and arrangements for accounting to the Board of Directors.*
   The policy should include a review procedure to take into account the entire organisation and signed off by the managing director or equivalent. The client should check that the health and safety policy is relevant to the nature and scale of the work undertaken by the principal contractor.

2. *What arrangements does the principal contractor have in place to manage health and safety, both at the head office and on site?*
   In particular there should be procedures explaining how the principal contractor shall discharge its duties under the Regulations. Evidence that the duties under the Regulations are communicated to the workforce should be expected.

3. *What resources and expertise does the principal contractor have for access to competent health and safety advice?*
   The health and safety advisor can be an employee, although there is no reason why such an advisor cannot be an independent organisation. The health and safety adviser must be able to provide general health and safety advice in addition to advice specific to construction health and safety issues.

4. *What arrangements does the principal contractor have for training employees to ensure that they have the skills and understanding required to comply with and discharge their duties under the Regulations?*
   The arrangements should include induction training, refresher training and general updating. Persons at all levels within the principal contractor should have appropriate training and information.

5. What are the individual training qualifications and experience of key staff who will be responsible for the control and supervision of construction work?

6. Does the principal contractor have a system for monitoring his health and safety procedures for implementing changes as a result of feedback from the monitoring procedures?

7. Does the principal contractor have an established means of consulting with its workforce on health and safety matters?

8. Does the principal contractor have records of all reportable events for at least the last 3 years?

9. Does the principal contractor have in place a system for reviewing all incidents and recording the action taken as a result including updating procedures?

10. What arrangements does the principal contractor have for appointing competent sub-contractors and consultants?

11. Is there a procedure capable of being audited to ensure that sub-contractors will also have arrangements for appointing competent contractors and sub-consultants?

12. What procedures does the principal contractor have in place for monitoring sub-contractor performance?

13. What convictions, if any, does the principal contractor have in respect of health and safety legislation over the last 3 years?

The post-qualification

1. By what means will the principal contractor co-operate and co-ordinate with the other parties identified in the project including procedures to involve the workforce in drawing up statements and systems of work?

2. What proposals does the principal contractor have to ensure that appropriate welfare facilities will be in place before persons start work on site?

3. What is the principal contractor's experience on similar projects?

4. What procedures will the principal contractor adopt for carrying out risk assessments and developing safe systems of work/ method statements for the hazards and risks identified by the client, designers, co-ordinator and other contractors?

5.   What procedures will the principal contractor adopt for preparing the construction phase plan for the particular project?

In addition to any enquiries as to competence a client may make of the principal contractor, the client can take the advice of the CDM co-ordinator who is required to provide such assistance by virtue of regulation 20(1)(a).

Requirements on and powers of the principal contractor

The requirements to which the principal contractor is subject and his powers are set out in regulations 22 and 23. Regulation 22 is the general duties of the principal contractor, in addition to its duties as a contractor as set out in regulations 13 and 19. Regulation 23 specifically deals with the principal contractor's duties with respect to the construction phase plan. The principal contractor is also required to comply with regulation 24, which deals with co-operation and consultation of works.

The fundamental duty of the principal contractor is at the head of the list of duties in regulation 22, which emphasises the management and co-ordination role of the principal contractor. For a contractor, other than the main or managing contractor, to be appointed as principal contractor would create practical contractual difficulties and is discouraged by the ACOP.

Regulation 22(1)(a) imposes the general duty upon the principal contractor to plan, manage and monitor the construction work as follows:

> *The principal contractor for a project shall plan, manage and monitor the construction phase in a way that ensures that, so far as is reasonably practicable, it is carried out without risks to health and safety.*

This general duty requires no explanation except that the Regulations go on to identify how the principal contractor is expected to undertake the general duty by including the facilitation in regulation 22(1)(a)(i) to (ii) as follows:

> *including facilitating*
>
> *(i)   co-operation and co-ordination between persons concerned in the project in pursuance of regulation 5 and 6;*
>
> *(ii) the application of the general principles of prevention in pursuance of regulation 7.*

The 1994 Regulations only required the principal contractor to take reasonable steps to ensure co-operation between all contractors. Under the Regulations, the principal contractor has to do whatever is necessary to plan, manage and monitor the construction phase in a way that ensures that, 'so far as is reasonably practicable' that there are no risks to health or safety. Regulation 3 of the management of Health and Safety at Work Regulations 1999 requires every contractor to prepare a risk assessment, which should address risks to employees and to other persons who may be affected by their activities. Therefore, the principal contractor should prepare a risk assessment as with every other contractor. Three aspects of the risk assessments prepared by the other contractors in a project will influence the role of principal contractor. These are:

1. The seriousness of the risk.

2. The nature of the assessment. The risk assessments prepared by other contractors are likely to comprise two parts. According to the normal activities of the contractor there is likely to be a generic risk assessment which will not require any amendment from project to project. However, for each individual project the second part of the risk assessment will need to be tailored to the particular risks for each project.

3. The principal contractor will be concerned to examine the inter-relationship between the risk assessments produced by the other contractors, including its own. Some activities of a contractor will have no effect on other contractors working on the project whereas in certain circumstances, activities of a contractor will affect one or more contractors to a greater or lesser extent.

The general principles are discussed in detail in Chapter 4. The discharge of the principal contractor's duties under regulation 22(1)(a) is made considerably easier than it might otherwise be by the mutual duties on all the other persons who will be subject to the principal contractor's planning, managing and monitoring.

The principal contractor has a duty to liaise with the CDM co-ordinator. This duty is set out in regulation 22(1)(b) as follows:

> *The principal contractor for a project shall liaise with the CDM co-ordinator in performing his duties in regulation 20(2)(d) during the construction phase in relation to any design or change to a design.*

In compliance with this duty, the principal contractor has to co-operate with the CDM co-ordinator and the designers to ensure that any change to a design during the construction phase is properly addressed, although the CDM co-ordinator has the primary role as provided for in regulation 20(2)(d) as follows:

> *the CDM co-ordinator shall take all reasonable steps to ensure co-operation between designers and the principal contractor during the construction phase in relation to any design or change to a design.*

The welfare arrangements and facilities on a site are ultimately the responsibility of the principal contractor as set out in regulation 22(1)(c), which provides:

> *The principal contractor for a project shall ensure that welfare facilities sufficient to comply with the requirements of Schedule 2 are provided with throughout the construction phase.*

The site rules are an important and essential means of regulating activity and behaviour on a site. Regulation 22(1)(d) requires that:

> *The principal contractor for a project shall where necessary for health and safety, draw up rules which are appropriate to the construction site and the activities on it (referred to in these Regulations as 'site rules').*

The principal contractor can only draw up and impose such site rules after considering the activities of all the other contractors on the site. The site rules must be adopted by and adhered to by all contractors and employees and workers on the site.

The site rules will cover all aspects of the construction work where there are risks, including access to parts of the site, traffic arrangements, hot work and so on. The site rules are the basis of communicating to the workers the means of avoiding risks.

The site rules should be in writing and the principal contractor should pay due attention to the languages and nationalities of the persons employed on the site. The site rules are important and therefore must be brought to the attention of all persons who are on the site. Importantly, the principal contractor is responsible for leasing and enforcing the site rules.

There should be a system of reporting breaches of site rules and all persons should be aware of the appropriate sanctions for such breaches, which at the most severe should include removal from the site and the project.

To enable the principal contractor to perform his duties it is essential that other contractors co-operate with the principal contractor. Regulation 22(1)(e) gives the principal contractor the authority to give instructions as follows:

> *The principal contractor for a project shall give reasonable directions to any contractor so far as is necessary to enable the principal contractor to comply with his duties under these Regulations.*

The directions which the principal contractor is authorised to give must be reasonable and only for the purposes connected with the principal contractor's own discharge of his duties. The contractors, who are required to comply with those directions, are not necessarily in any contractual relationship with the principal contractor, but are bound by the commensurate duty to comply with such instructions by virtue of regulation 19(2)(c).

The principal contractor's role of co-ordination, as set out in regulation 22(1)(a)(i), is highlighted by the fact that a principal contractor has to communicate with the other contractors for the purposes of planning and preparation of each contractor's work. This duty is set out in regulation 22(1)(f) as follows:

> *The principal contractor for a project shall ensure that every contractor is informed of the amount of time which will be allowed to him for planning and preparation before he begins construction work.*

It is apparent that, to discharge this duty, the principal contractor has to have an overall programme/work schedule for the project which must take into account and allow for planning and preparation for each contractor. The time that is to be allowed does not have to be reasonable although not to give reasonable time to a contractor will probably be in breach of the general principles of prevention as required by regulation 22(1)(a)(ii).

The principal contractor's role in preparing a construction phase plan cannot be completed without consulting with the other contractors. This is the essential role of communication and to achieve the co-operation and co-ordination as required by regulation 22(1)(a). The duty is expressly set out in regulation 22(1)(g) as follows:

> *The principal contractor for a project shall where necessary, consult a contractor before finalising such part of the construction phase plan as is relevant to the work to be performed by him.*

It is not necessary for the principal contractor to disclose and discuss the entire construction phase plan with a contractor but a principal contractor should be wary of withholding any information unless it is sure that other activities on the site would not impinge in any way on a contractor's own work.

Once the construction phase plan has been finalised, the principal contractor is required to give access to the construction phase plan to every contractor. If the principal contractor has complied with regulation 22(1)(g), there should be no surprises for a contractor in a finalised construction phased plan. The duty which applies to the principal contractor is set out in regulation 22(1)(h):

> *The principal contractor for a project shall ensure that every contractor is given, before he begins construction work and in sufficient time to enable him to prepare properly for that work, access to such part of the construction phase plan as is relevant to the work to be performed by him.*

The principal contractor has to provide the construction safety plan in sufficient time and with sufficient detail. It would be difficult for the principal contractor to gauge what sufficient time or sufficient detail is without having obtained the fullest co-operation from a contractor as to how it may be relevant to the contractor's work and the work of other contractors.

The final element of the duties of the principal contractor with respect to co-ordinating the input from the other contractors is set out in regulation 22(1)(i):

> *The principal contractor before a project shall ensure that every contractor is given, before he begins construction work and in sufficient time to enable him to prepare properly that work, such further information as he needs –*
>
> *(i)   to comply punctually with the duty under regulation 13(7); and*
>
> *(ii) to carry out the work to be performed by him without risk so far as is reasonably practicable, to the health and safety of any person.*

The sufficiency of time to enable a contractor to prepare for his work links back to regulation 22(1)(f) pursuant to which the principal contractor must inform every contractor of the amount of time allowed to him for planning and preparation.

Regulation 13(4) is the duty of a contractor to ensure that the site is secure against access by unauthorised persons to the site. The contractor should have the opportunity to obtain any further information he requires, having taken account of the contents of the construction phase plan, so that his works can be varied or modified to further mitigate any risk to the health and safety of any person.

The principal contractor has a duty to provide the information to the CDM co-ordinator that is required for the health and safety file. The principal contractor is an important conduit in the flow of information from the various contractors through to the CDM co-ordinator and to the health and safety file. The importance of this role is set out as an express duty in regulation 22(1)(j) which provides:

> *The principal contractor for a project shall identify to each contractor the information relating to the contractor's activity required by the CDM co-ordinator for inclusion within the health and safety file in pursuance of regulation 20(2)(e) and ensure that such information is promptly provided to the CDM co-ordinator.*

The principal contractor is entitled to expect some leadership and overall project management of this role from the CDM co-ordinator by virtue of regulation 20(2)(d) but it is likely to be the result of contractual obligations with its sub-contractors that it can press for such information without delay. The principal contractor has only the authority of the Regulations when ensuring that information is provided by contractors other than those it is in some contractual relationship with.

The notice given to the health and safety executive pursuant to regulation 21 has to be displayed on a site so that all persons entering upon a site are able to read the notice and understand the status of the site for the management of health and safety.

This duty is set out in regulation 22(1)(k) which provides:

> *The principal contractor for a project shall ensure that the particulars required to be in the notice given under regulation 21 are displayed in a readable condition in a position where they can be read by any worker engaged in the construction work.*

The particulars would reasonably form an appendix to the site rules and such particulars should be included in the welfare facilities, site offices and in common positions around the site.

All the contractors on a site have a responsibility to prevent access by unauthorised persons to the construction site. However, the principal contractor is given the overall responsibility thus breaching any gaps in liability or effectiveness there may be as between the other contractors. This overwriting duty is set out in regulation 22(1)(l) as follows:

> *The principal contractor for a project should take reasonable steps to prevent access by unauthorised persons to the construction site.*

What are reasonable steps would be a question of fact which will vary from site to site. It is also not reasonable for the principal contractor to rely entirely upon the other contractors working on the site to have discharged the principal contractor's duty by default.

The principal contractor has the important role of man management and this means that workers should be properly informed and trained.

This important duty is set out in regulation 22(2) which provides:

> *The principal contractor shall take all reasonable steps to ensure that every worker carrying out the construction work is provided with –*
>
> *(a) a suitable site induction;*
>
> *(b) the information and training referred to in regulation 13(3) by a contractor on whom a duty is placed by that regulation; and*
>
> *(c) any further information or training which he needs for the particular work to be carried out without undue risk to health or safety.*

Unless the principal contractor takes it upon itself to provide the site induction, it should require evidence from the contractors that workers have had the benefit of a site induction. It is not good enough for site induction to take place after workers have started work. Although the Regulations do not provide that the site induction should be delivered before commencing construction work it is implicit in so far as the definition of induct is 'to introduce'.

The principal contractor can only ensure that regulation 22(2) is satisfied if it maintains a regular system of monitoring as to the deployment of workers employed by other contractors. For large construction sites, the principal contractor should consider maintaining a central register of all workers on the site, identifying the particular work that each worker is employed to undertake and that the site induction, relevant information and training have been delivered. The principal contractor cannot raise as a defence that it was the responsibility of a contractor to have provided such training for his own workers when in fact no such induction or training took place.

Co-operation and consultation with workers

The Regulations create a duty on all persons concerned on a project to seek co-operation and co-operate with others for the purposes of filling the duties and functions of the Regulations by virtue of regulation 5. However, the principal contractor is given an additional duty which reflects the importance of the need to initiate and promote co-operation of the workers together with express duty to consult with workers. It is noteworthy that there is no express equivalent duty on contractors where the project is non-notifiable, although the ACOP can be construed as imposing a duty to consult with their workers in any event. This is implied, presumably, as part of the general obligation under regulation

13(2) to manage the construction work so that it is carried out with no risks to health and safety, so far as is reasonably practicable. There is also the obligation to train workers, the scope of which will depend upon their existing skills and experience, the extent of which can only be ascertained by consulting with the workers.

In the absence of any express obligation on contractors to consult with their employees or workers, but with the clear views set out in the ACOP, contractors would be well advised to consult with their employees and workers on non-notifiable projects. The remainder of this section deals with the express duties upon the principal contractor.

Co-operation
The principal contractor has a duty to make and maintain arrangements for co-operation as required by regulation 24(a) which states:

> *The principal contractor shall make and maintain arrangements which will enable him and the workers engaged in the construction work to co-operate effectively in promoting and developing measures to ensure the health, safety and welfare of the workers and in checking the effectiveness of such measures.*

The arrangements to be made by the principal contractor go beyond the need for co-operation with his own employees and workers. The arrangements will cover all persons on the site including the other contractors, client representatives, designers and the CDM co-ordinator.

The extent and nature of the arrangements will vary according to the nature of each project. Irrespective of the particular aspects of the project, each of the other contractors, client representatives, designers and CDM co-ordinators are required to assist and actively promote co-operation for and on behalf of their own employees and workers.

The reference within regulation 24(a) to the principal contractor and the workers clearly points to arrangements which will allow for collaboration and a mutual exchange of ideas and information. As the ACOP states, it involves a joint commitment to solving problems together.

The principal contractor cannot afford to leave the arrangements for co-operation to chance. The arrangements should be planned and recorded in writing and, moreover, represent the outcome of consultation which is also required by regulation 24(b).

The planning of the arrangements must take account of the size and complexity of the construction work, the extent of hazards and the associated risks and the number of workers and contractual arrangements.

The arrangements must be effective and it is the principal contractor's duty to monitor the arrangements to ensure that they are effective. The danger in making arrangements at the start of a project is that they simply become another routine box-ticking exercise supported by regular meetings that are to all intents and purposes sterile and a contractually motivated exchange.

Ultimately, the client cannot stand by if he has reason to believe that the arrangements are not effective, especially if so advised by the CDM co-ordinator. The client's duty, as set out in regulation 9(1)(a), should be manifested through the construction contract with the principal contractor.

Consultation

The principal contractor's duty to consult extends to all workers involved on a project. The duty is set out in regulation 24(b) as follows:

> *The principal contractor shall consult those workers or their representatives in good time on matters connected with the project which may affect their health, safety or welfare so far as they or their representatives are not so consulted on those matters by any employer of theirs.*

Consultation implies a dialogue which is opened by the principal contractor. The principal contractor has to listen and take account of what workers say, at all stages of the construction work and all aspects of the health and safety environment. On large sites, where the principal contractor may have a substantial number of employees, the consultation can be done through worker representatives. In the case of sites where there is trade union representation, the worker representatives may be appointed by the recognised trade unions under the Safety Representatives and Safety Committees Regulations 1977. Alternatively, the workforce can nominate and elect their own representatives under the Health and Safety (Consultation with Employees) Regulations 1996. While it may be appropriate to consult with representatives, no worker should ever be prevented from having an opportunity to make his or her own representations. There is a natural resistance for workers to report defects or examples of poor management and the principal contractor should be sympathetic to making arrangements for confidential consultation if appropriate. Employees now have the benefit of protection in making such representations by the protected disclosure provisions of the Employment Rights Act 1996 (as amended).

Information

The identification of hazards, assessment of risks, method statements and systems of work, management–worker co-operation and consultation, notices, changes to the project and accident reporting all generates information. All persons on a site are entitled to such information by virtue of regulation 24(c) which states:

> *The principal contractor shall ensure that such workers or their representatives can inspect or take copies of any information which the principal contractor has or which these Regulations require to be provided to him which relates to the planning and management of the project, or which otherwise might affect their health, safety or welfare at the site.*

The principal contractor's administrative burden should not be underestimated and the need for an effective document management system cannot be overstated. A principal contractor could be severely criticised in the event that such information was not provided in good time.

There are some exclusions from categories of information which a principal contractor has to make available and these are set out as part of Regulation 24(c) as follows:

except any information –

(i) *the disclosure of which would be against the interests of national security;*

(ii) *which he could not disclose without contravening a prohibition imposed by or under an enactment; or*

(iii) *relating specifically to an individual, unless he has consented to it being disclosed; or*

(iv) *the disclosure of which would, for reasons other than its effect on health, safety or welfare at work, cause substantial injury to his undertaking or, where the information was supplied to him by some other person, to the undertaking of that person; or*

(v) *obtained by him for the purpose of being, prosecuting or defending any legal proceedings.*

The principal contractor's responsibility for the construction phase plan

The principal contractor has the primary responsibility for the construction phase plan the general duties of which are set out in regulation 23.

Chapter 13 is devoted specifically to the construction phase plan where the principal contractor's role is examined in detail.

Checklist for principal contractor

The duties applicable to contractors for all construction work and notifiable projects apply to the principal contractor. Listed below are the additional duties of the principal contractor for notifiable projects.

Notifiable projects	Regulation
1. Are you aware of your duties as principal contractor under the Regulations?	22, 23 and 24
2. Have you confirmed the identity of the CDM co-ordinator?	22(1)(b)
3. Have you received a copy of the notification of the project to the Executive?	19(1)(c)
4. Are you competent to undertake the duties of principal contractor in dealing with health and safety issues involved in the management of the construction phase?	4(1)(b)
5. Have you taken steps to ensure that the construction phase is properly planned, managed and monitored, with adequately resourced competent site management appropriate to the risk and activity?	22(1)(a)
6. Have you provided to every contractor who will work on the project the information for the amount of time they will be allowed for planning and preparation before the start of construction work?	22(1)(f)
7. Have you provided to every contractor who will work on the site, the information about the project that they will need for planning their own construction work without risk to health and safety?	22(1)(i)
8. Have you established a communication system to facilitate co-operation between the contractors and co-ordination of their respective construction works to ensure safe working?	22(1)(a)(i)
9. Have you taken steps to prepare a construction phase plan which has been developed in discussion with, and communicated to, contractors whose own construction work will be affected by it?	22(1)(g)

Notifiable projects		Regulation
10.	Have you ensured that the construction phase plan is completed before construction work begins?	23(1)(a)
11.	Have you a management system in place to ensure that the construction phase plan is implemented and reviewed and updated as the project progresses?	23(1)(b)
12.	Have you taken reasonable steps to prevent unauthorised access to the site?	22(1)(l)
13.	Have you ensured that there will be suitable welfare facilities for all construction workers on the site at the start of the construction phase?	22(1)(c)
14.	Have you satisfied yourself that the contractors and designers you engage are competent and adequately resourced?	4(1)(a)
15.	Have you published and circulated the necessary site rules together with the means of policing and enforcing such rules?	22(1)(d)
16.	Have you provided access to the relevant part of the construction phase plan to contractors in sufficient time for them to plan their work?	22(1)(h)
17.	Have you liaised with the CDM co-ordinator on design carried out and/or completed during the construction phase and considered its implications on the construction phase plan?	22(1)(b)
18.	Have you provided the CDM co-ordinator with the information necessary and relevant to the health and safety file?	22(1)(j)
19.	Have you satisfied yourself that all workers on a construction site have been provided with suitable health and safety induction, information and training?	22(2)
20.	Have you taken steps to ensure that the workforce is consulted about health and safety matters and implement procedures to facilitate co-operation?	24
21.	Have you displayed the project notification to the executive?	22(1)(k)

12 The contractor

Definition
Selecting a contractor
Duties of contractors
   Sub-contractors
   Information and training
   Visitors to site
   Welfare facilities
Additional duties of contractors
Basic checklist of considerations for contractors

Definition

A contractor is defined in regulation 2(1) as:

> *any person (including a client, principal contractor or other person referred to in these Regulations) who, in the course or furtherance of a business carries out or manages construction work.*

A contractor is concerned with construction work, the definition and scope of which has been reviewed in Chapter 3.

The definition refers to carrying out or managing construction work. A contractor does not have to be engaged in both activities such that a contractor can be engaged solely in managing construction work provided that it is in the course or furtherance of a business.

The definition is a sweep up of all persons involved in construction to ensure that they are all subject to the Regulations. The client and principal contractor are expressly included in the definition, whereas the 'other person' would include CDM co-ordinator and designer as being referred to in the Regulations. Sub-contractors and suppliers are all deemed to be contractors if they are involved to any extent in managing or carrying out construction work.

Selecting a contractor

By virtue of regulation 4, any person arranging for a contractor to carry out or manage construction work has to have taken reasonable steps to ensure that the contractor is competent.

Matters which a person should consider before appointing a contractor in accordance with regulation 4(1)(a) are essentially the same as those matters contained in the suggested checklist for the principal contractor in Chapter 10.

Duties of contractors

The duties of contractors set out in regulation 13 apply to all projects whether they are notifiable or not.

Regulation 13(1) imposes on a contractor the duty to enquire that the client for a particular project is aware of his duties under the Regulations as follows:

> *No contractor shall carry out construction work in relation to a project unless any client for the project is aware of his duties under these Regulations.*

By the contractor complying with this duty, the client will become aware of the need to comply with the Regulations. On a large project, where the contractor may be at the bottom of a long contract chain, the contractor can be satisfied that the client is aware of his duties if the appointment of the principal contractor has been drawn to the attention of the sub-contractor. On small projects, however, the situation may not be quite so obvious. The small project may not be notifiable and a contractor acting as sub-contractor may have suspicions that its employing contractor is unaware of, or is complying with, the Regulations and further compounded by the fact that the client is also unaware of the Regulations. In those circumstances, irrespective of the contractual relationships and obligations, the sub-contractor should make it his or her business to be satisfied that the client is aware of his or her duties under the Regulations.

The duty of a contractor to ensure that his work is carried out without risk to health and safety is expressed in regulation 13(2) which states:

> *Every contractor shall plan, manage and monitor construction work carried out by him or under his control in a way which ensures that, so far as is reasonably practicable, it is carried out without risk to health and safety.*

The contractor's duties fall into three activities, namely planning, managing and monitoring construction work. These activities apply to construction work carried out directly by the contractor and work carried out under his control. Work carried out under a contractor's control is usually under a contractual arrangement where the extent of control is expressly understood. However, the extent of control exercised by a contractor may not be so clearly understood, especially if the client has arranged for a specialist, as an example, to work on the site in close quarters with the contractor in circumstances where the specialist work will become an integral part of the contractor's own works.

The duty of the contractor to ensure the construction work is carried out without risks to health and safety is subject to the test of being reasonably practicable. The test of reasonably practicable recognises the inevitable: that construction work can rarely be carried out without some risk, however small, to health and safety.

Sub-contractors

There will be many instances where a contractor will appoint sub-contractors, in which case the contractor will have to ensure that the sub-contractor complies with the Regulations and the contractor's own actions for appointment comply with the Regulations. Regulation 13(3) requires that:

> *Every contractor shall ensure that any contractor whom he appoints or engages in his term in connection with a project is informed of the minimum amount of time which will be allowed to him for planning and preparation before he begins construction work.*

The contractor is required to provide information of the minimum amount of time for tendering before construction work begins. The contractor will be constrained by the time which he or she has been given in which to prepare before the start of the construction work. The importance of this requirement is to provide an opportunity for sub-contractors to assess whether that is sufficient time for their own purposes in planning the work. In those situations where a sub-contractor is of the view that he or she cannot plan and prepare the work in accordance with the Regulations within the time allocated to say that they will not accept an appointment. If the sub-contractor accepts the appointment he or she will be impliedly acknowledging that the planning and preparation for the construction work and allocation of resources can be adequately provided for within the allotted time.

Information and training

The contractor's duties to its workers and those of other contractors or self-employed tradesmen subject to the contractor's control are set out in regulation 13(4). The emphasis is upon training its employees and other workers and ensuring that there is an effective means of communicating relevant information. Regulation 13(4) states:

> *Every contractor shall provide every worker carrying out the construction work under his control with any information and training which he needs for the particular work to be carried out safely and without risk to health, including. . . .*

The contractor is under an obligation to have provided appropriate training for his or her own employees and it is also his or her responsibility to ensure that any sub-contractors or self-employed persons have received the relevant training. It is not necessary for the training to ensure that all the workers become competent, although it is essential that persons responsible for supervising others are competent. The training requirements are set out in regulations 13(3)(a)–(f), although it should be noted that the points for training and communication are not exhaustive.

Regulations 13(4)(a)–(f) can be either treated as an agenda for a face-to-face meeting with the employees or workers, or alternatively can be provided in written form. Contractors should not assume that employees are sufficiently literate to be able to understand written instructions and ideally, therefore, the training should be a combination of a face-to-face presentation and written materials which the employee or worker can retain.

Regulation 13(4)(a) requires the contractor to provide:

> *suitable site induction, where not provided by any principal contractor.*

What should be included in a site induction session which is not included in the remaining items at (b)–(f) is not obvious. However, it would be a good idea to provide background information on the construction works. This would include the name of the project, the purpose of the completed project, the programme for construction and completion, basic statistics such as number of workers on the site, before progressing to the more prescriptive requirements at (b)–(f).

The contractor is required to provide to every worker carrying our construction work under his control:

> *(b) information on the risks to their health and safety –*

*(i)   identified by his risk assessment under regulation 3 of the
        Management of Health and Safety at Work Regulations
        1999; or*

*(ii) arising out of the conduct by another contractor of his under-
        taking and of which he is or ought reasonably to be aware.*

It is axiomatic that a contractor cannot comply with regulation
13(4)(b) unless he or she has carried out the risk assessments under
regulation 3 of the Management Regulations and communicated those
to the workers and employees. The conduct of another contractor in
carrying out its construction work is unlikely to be known to the contractor unless it has been informed by a principal contractor or designer, or
the client. In addition to the information provided to the contractor, he
or she would also be able to see the activities of another contractor. Even
if another contractor is working on the other side of the site, that
contractor's activities may still represent risks for the contractor's own
workers and employees – in particular, traffic movement, laying of
power cables, etc.

The contractor is required to monitor the risk assessment and to the
extent that there is any change in those risks communicate them without
delay to the workers and employees.

The outcome of undertaking risk assessments is the planning of
measures to minimise the risks. Regulation 13(4)(c) provides the relevant
duty as follows:

> *the measures which have been identified by the contractor in consequence
> of the risk assessment as the measures he needs to take to comply with the
> requirements and prohibitions imposed upon him by or under the relevant
> statutory provisions.*

The measures that have been identified by the contractor are part of
the planning stage. The measures that a contractor should consider
would include physical protection, confining certain activities to different areas, management of movement and even shift patterns and
timing of tea breaks, etc.

On any site to which the Regulations apply there will be site rules, and
paragraph 13(4)(d) imposes a duty on the contractor to provide each
worker or employee with site rules. It would be advisable for the contractor to supply to each worker and employee a copy of the site rules that
can be easily carried in a pocket or tucked away in a protective helmet.
Notices displaying the site rules should be displayed at prominent points
around a site both on the constructions works and in the welfare
facilities.

A contractor is required to have planned for procedures in the event of any incidents by virtue of regulation 13(4)(e) which refers to:

> *the procedures to be followed in the event of a serious and imminent danger to such workers.*

Serious and imminent danger means that the risk to health and safety is serious and that the risk arises as soon as work starts or immediately if the work is in progress. The planning for evacuation to a safe place under such circumstances is an essential part of planning for health and safety. On sites where the construction work is non-notifiable, the client and the contractor should ensure that the procedures apply to the whole site so as to avoid confusion or panic in the event of a serious and imminent danger threatening the site.

Regulation 8 of the Management Regulations 1999 requires employers to establish and give effect to appropriate procedures to be followed in the event of serious and imminent danger to persons at work as follows:

> *The contractor is bound by regulation of the Management Regulations to nominate a sufficient number of competent persons to implement the procedures in accordance with regulation 8(1)(b) of the Management Regulations.*

The workers and employees need to be aware of the identity of persons responsible for the procedures where there is a serious or imminent danger as required by regulation 13(4)(f) which specifically imposes the duty to provide every worker with:

> *the identity of the persons nominated to implement those procedures.*

Workers and employees should be informed of the identity of fire marshalls, security staff, first aiders and any other persons, all of whom should have been appropriately trained to meet the requirements of competence under the Management Regulations.

Regulation 13(4) requires the contractor to provide information and training, whereas regulation 13(5) requires the contractor to provide his employees with health and safety training as follows:

> *Without prejudice to paragraph (4), every contractor shall in the case of any of his employees provide those employees with any health and safety training which he is required to provide to them in respect of the construction work by virtue of regulation 13(2)(b) of the Management of Health and Safety at Work Regulations 1999.*

The health and safety training referred to in regulation 13(2)(b) of the Management Regulations has to be adequate for employees being exposed to new or increased risks. Regulation 13(2)(b) of the Management Regulations lists the circumstances in which new or increased risks can occur as follows:

> *(i)   their being transferred or given a change of responsibilities within the employer's undertaking,*
>
> *(ii) the introduction of new work equipment into or a change respecting work equipment already in use within the employer's undertaking,*
>
> *(iii) the introduction of new technology into the employer's undertaking, or*
>
> *(iv) the introduction of a new system of work into or a change respecting the system of work already in use within the Employer's undertaking.*

The contractor has to keep under review the health and safety training status of his employees and consider in the event of new equipment or changes in systems of work as to whether his employees need further training.

Visitors to site

Protection of the public and unauthorised visitors to the site is an important responsibility of a contractor. Regulation 13(6) provides:

> *No contractor shall begin work on a construction site unless reasonable steps have been taken to prevent access by unauthorised persons to that site.*

The perimeter and delineation of a construction site needs to be understood by the contractor. There is the construction site for the project and within the overall construction site there are likely to be discrete areas or zones.

Unauthorised persons will include members of the public, but will also include other workers or employees on the construction site who do not have the necessary permits or competence to enter upon a part of the construction site where particularly hazardous operations may be in progress.

A contractor, therefore, has to consider the security of the entire site and if, for good reasons, the contractor is unsatisfied that the site is secure against unauthorised persons, the contractor should not begin work. It will be a bold contractor who, not being the principal contractor, or the main contractor, refuses to begin work because the site is not secured against unauthorised persons. However, any claim brought by a principal contractor or main contractor against such a contractor for delay could be defended on the grounds that the principal contractor or main contractor was in breach of its statutory duty and thus would have put the contractor in breach of its statutory duties.

Where a contractor is to undertake its own work within a defined area or zone, it should take steps to ensure that only authorised persons are permitted to enter.

As to whether the steps to prevent unauthorised persons to enter upon a site are reasonable or not will depend upon the potential hazards and risks. Where the hazard and risks have been assessed as very low, the reasonable steps may consist merely of some temporary barrier. Where the hazard and risks have been assessed to be high, the barrier should be of such a design as to be continuous and secure against determined action to enter. The physical barrier should also, in a high-risk environment, be kept under surveillance by CCTV or patrolling security staff.

The risk to a contractor of an unauthorised person entering upon a site and suffering harm is of a potential prosecution under the Regulations and a claim from such an injured person under the Occupiers Liability Acts 1957 and 1984.

The onus on the contractor to satisfy itself that the construction site is secure to the appropriate extent proportionate to the hazard and risks cannot be overstated.

Welfare facilities

Every contractor should provide welfare facilities for its employees or any other worker under his control. Regulation 13(7) states:

> *Every contractor shall ensure, so far as is reasonably practicable, that the requirements of Schedule 2 are complied with throughout the construction phase in respect of any person at work who is under his control.*

The contractor can make arrangements for the use of the welfare facilities provided by another contractor on the construction site or even provide such facilities outside of the construction site. A contractor may not be able to achieve full compliance with Schedule 2 because it is not reasonably practicable to do so, but there cannot be any defence for a contractor who fails to achieve the minimum compliance with Schedule 2 commensurate with the basic requirements of public health and public order.

Additional duties of contractors

In circumstances where the construction work is notifiable, a contractor is subject to additional duties. Regulation 19(1) states:

> *Where a project is notifiable, no contractor shall carry out construction work in relation to the project unless –*

Thus, the contractor has additional duties which are mandatory. The remaining provisions of regulation 19 list the additional duties which reflect the mandatory appointments of the CDM co-ordinator and principal contractor. This is highlighted by regulation 19(1)(a) which states that a contractor shall not carry out the construction work unless:

> *he has been provided with the names of the CDM co-ordinator and principal contractor.*

The contractor shall not start construction work until he has sight of the relevant part of the construction phase plan as provided for by regulation 19(1)(b) as follows:

> *he has been given access to such part of the construction phase plan as is relevant to the work to be performed by him, containing sufficient detail in relation to such work.*

The contractor should consider the relevant part of the construction phase plan critically and assess whether he or she should have access to other parts of the construction phase plan. To accept without question the relevant part, as assessed by the principal contractor or co-ordinator, is to deny the fact that the contractor probably has a better knowledge and understanding of the work he or she is to undertake.

Finally, before the contractor should start construction work, he or she has to be satisfied that the appropriate notice has been sent to the Executive as required by regulation 19(1)(c) as follows:

> *notice of the project has been given to the Executive, or as the case may be the Office of Rail Regulation, under regulation 21.*

It is only when the three crucial pieces of information at regulation 19(1) have been provided to the contractor that the contractor is able to start work. To start construction work without the information not only puts the contractor in immediate breach of regulation 19(1) but also puts the contractor, its employees and workers under its control at unforeseen and undetermined risks.

The principal contractor has the duty to plan, manage and monitor the construction phase which includes co-operation and co-ordination between contractors and itself by virtue of regulation 22(1)(a). The reciprocal duties of contractors to co-operate with and supply information to the principal contractor are set out in regulation 19(2). Regulation 19(2)(a) is the important duty to provide information to the principal contractor about the specific health and safety issues arising from the contractor's own construction work as follows:

> *Every contractor shall promptly provide the principal contractor with any information (including any relevant part of any risk assessment in its possession or control) which –*
>
> *(i) might affect the health and safety of any person carrying out the construction work or of any person who might be affected by it,*
>
> *(ii) might justify a review of the construction phase plan.*

The importance of the information to be provided by the contractor is underlined by the recognition that such information might justify a review of the construction phase plan. Information should be provided that might affect the health or safety of any person at work carrying out the construction work or by any person who might be affected by it. This will include the contractor's own employees and workers under its control and all other workers and other persons named in the Regulations who are likely at any time to be on site as authorised. The information to be provided by the contractor should include the risk assessments, a detailed description of the works he would be undertaking and the operational details such as the identification of plant and equipment and time of working.

The information should be provided promptly rather than as soon as reasonably practicable. It must be implicit that the information should be provided before any operations start which represent a risk to health and safety and may not have been factored into the principal contractor's own assessments and any need to review the construction phase plan.

The primacy of the principal contractor with respect to health and safety is reinforced by virtue of regulation 19(2)(c) which provides:

> *Every contractor shall comply with any directions of the principal contractor given to him under regulation 22(1)(e).*

Provided that directions of the principal contractor are limited to enabling him or her to comply with his duties under the Regulations, a contractor is bound, by virtue of this regulation, to comply with such directions. This will be so regardless of any contractual relationship or otherwise, which might exist between the principal contractor and contractor.

Regulation 19(2)(d) provides:

> *Every contractor shall promptly provide the principal contractor with the information in relation to any death, injury, condition or dangerous occurrence which the contractor is required to notify or report under the Reporting of Injuries, Diseases and Dangerous Occurrences Regulations 1995.*

The information which the contractor is required to compile and notify or report to the enforcing authorities, by virtue of the reporting of Injuries, Diseases and Dangerous Occurrences Regulations 1995, should be promptly provided by delivery of copies of such information to the principal contractor. The principal contractor needs the information to monitor compliance with health and safety law and ensure that the arrangements for the management of health and safety remain appropriate throughout the duration of the construction phase.

No project can be treated as entirely completed until the health and safety file has been compiled. All persons engaged in a project have a responsibility to contribute to the health and safety file including the contractor. Regulation 19(2)(a)(iii) provides:

> *Every contractor shall promptly provide to the principal contractor with any information (including any relevant part of any risk assessment in his possession or control) which he has identified for inclusion in the health and safety file in pursuance of regulation 22(1)(j).*

The information that the contractor has to provide should be accurate and adequate. The contractor should realise that, in providing this information, it will be relied upon by the owner of the project and any other persons involved in later projects that might include the construction work undertaken by the contractor. The requirement to provide the information promptly will often be a contractual term such that the principal contractor is not in breach of its obligations to the client.

The additional duties of contractors arise only where the project is notifiable in which case there will be a construction phase plan. Contractors have duties which arise from the creation and existence of the construction phase plan which are set out in regulation 19(3).

The contractor's duty to plan, manage and monitor construction work carried out by him or her, or under this control pursuant to regulation 13(2), is extended by virtue of regulation 19(3)(a), which provides:

> *Every contractor shall in complying with his duty under regulation 13(2) take all reasonable steps to ensure that the construction work is carried out in accordance with the construction phase plan.*

The construction work referred to in this regulation is the same construction work referred to in regulation 13(2). That is to say, that the contractor's duty is only with respect to the construction work carried out by him or her, or under his or her control. It also follows, that regulation 19(1)(b) has been satisfied insofar as the contractor has been given access to the relevant parts of the construction phase plan.

There will be circumstances where a contractor will take the view, having examined the construction phase plan, that he or she is unable to comply with aspects of the construction phase plan. This might arise where the contributors to the construction phase plan up to that stage had not fully appreciated the risks to health and safety of construction work to be undertaken by the contractor. In those circumstances, regulation 19(3)(b) provides:

> *Every contractor shall take appropriate action to ensure health and safety where it is not possible to comply with a construction phase plan in any particular case.*

The contractor is obligated to take 'appropriate action', although the regulation does not expressly state that appropriate action should ensure that the construction work does comply with the construction phase plan. In the first instance, it will require the contractor to plan, execute and manage its construction work so that it does comply with the construction phase plan. This may be the cause of a contractual dispute if the contractor is able to demonstrate that the basis for its plan for executing and managing the work would have been different if it had been aware of the full contents of the construction phase plan.

If the appropriate action entails changes to the execution and managing of the construction work that will impinge upon the activities and other aspects of the construction phase plan, the contractor is obliged by virtue of regulation 19(3)(c) as follows:

> *Every contractor shall notify the principal contractor of any significant finding which requires the construction phase plan to be altered or added to.*

Thus, if the contractor cannot take appropriate action without having an impact on others which has not been foreseen by the construction phase plan, the contractor has to make this known to the principal contractor. The principal contractor cannot ignore such a notice from a contractor without taking a considerable risk with the health and safety of employees and workers of other contractors and for its own liability under the Regulations.

Basic checklist of considerations for contractors

All projects	**Regulation**
1. Have you ascertained that the client is aware of his duties?	13(1)
2. Are you satisfied that you have the competence to accept the appointment or engagement as a contractor?	4(1)(b)
3. Are you satisfied that your employees and workers involved in managing design or construction work are competent, or will be working under the supervision of a competent person?	4(1)(c)
4. Have you taken steps in planning, managing and monitoring the construction work to make sure that workers under your control are safe from the start of their work on site?	13(2)
5. Have you satisfied yourself that any contractor who you appoint or engage to work on a project is informed of the minimum amount of time which will be allowed for them to plan and prepare before starting work on site?	13(3)
6. Have you provided workers under your control, including self-employed workers, with any necessary information, including relevant aspects of work being undertaken by others?	13(4)
7. Have you provided health and safety training to employees?	13(5)
Have you taken reasonable steps to prevent unauthorised access to the site?	13(6)
8. Have you ensured that any design work undertaken by you complies with regulation 11?	11
9. Have you planned for and provided the welfare facilities in Schedule 2?	13(7)
10. Have you taken steps to comply with part 4 of the Regulations as appropriate?	25(1)
11. Do you have a plan or established means of communication to facilitate co-operation with others and co-ordination of your work with the work of others?	5 and 6

All projects	Regulation
12. Have you considered the need to obtain specialist advice where excavations and cofferdams or caissons are involved?	31 and 32
Notifiable projects	
13. Have you received a copy of the notification of the project to the Executive?	19(1)(c)
14. Do you know the identity of the CDM co-ordinator and principal co-ordinator?	19(1)(a)
15. Have you made contact with the principal contractor, CDM co-ordinator, other contractors and designers working on the project or adjacent sites for the purposes of establishing co-operation co-ordination?	5 and 6
16. Have you informed the principal contractor about risks to others that will be created by your construction work?	19(2)(a)
17. Have you provided details to the principal contractor of any sub-contractor or self-employed person who you have engaged to carry out the construction work?	19(2)(b)
18. Have you received a copy of the construction phase plan, or a relevant part of the construction phase plan?	19(1)(b)
Have you taken reasonable steps to ensure that the construction work is carried out in accordance with the construction phase plan?	19(3)(a)
Where it is not possible to comply with the construction phase plan have you taken action to ensure health and safety?	19(3)(b)
19. Have you checked the construction phase plan to identify any risks created by others which will have significant implications for your construction work or for the management of the project?	19(3)(c)
20. Have you established communication with the principal contractor for the purposes of receiving reasonable directions and providing feedback on the construction phase plan?	19(2)(c)

All projects	Regulation
21. Have you put in place established rules for communicating accidents and dangerous occurrences to the principal contractor?	19(2)(d)
22. Have you collected together information for the health and safety file to give to the principal contractor?	19(2)(a)(iii)

13 The construction phase plan

Definition
Background
The duty to prepare the construction phase plan
Checklist for content of construction phase plan

Definition

The construction phase plan is defined in regulation 2(1) as meaning:

a document recording the health and safety arrangements, site rules and any special measures for construction work.

The construction phase plan is a document and therefore must be kept in a form that is reproducible and secure as required by regulation 2(2) which provides:

Any reference in these Regulations to a plan, rules, document, report or copy includes the plan, rules, document, report or copy which is kept in a form –

(a) in which it is capable of being reproduced as a printed copy when required;

(b) which is secure from loss or unauthorised interference.

The plan is only required to take account of the construction phase which is defined in regulation 2(1) as meaning:

the period of time starting when construction work in any project starts and ending when construction work in that project is completed.

Therefore, the construction phase should not commence until a construction phase plan has been prepared. It is the responsibility of the client to ensure that the construction phase does not start until the construction phase plan has been completed in accordance with regulation 16 as follows:

Where the project is notifiable, the client shall ensure that the construction phase does not start unless –

  (a) the principal contractor has prepared a construction phase plan which complies with regulations 23(1)(a) and 23(2); and

  (b) he is satisfied that the requirements of regulation 22(1)(c) (provision of welfare facilities) will be complied with during the construction phase.

Accidents associated with construction work cannot occur on a project until construction work commences. This duty on the client to prevent the start of construction work until the construction phase plan has been prepared is arguably the most important duty of the client in exerting influence over the management of health and safety during construction work.

Background

The construction phase plan replaces the health and safety plan required by the 1994 Regulations. It no longer represents the continuous thread throughout the project as a management tool, starting with design in the pre-construction phase. Its importance however, remains as central to the Regulations as did the health and safety plan to the 1994 Regulations.

The principal contractor takes the primary responsibility for the management of health and safety on a construction site, and the Regulations ensure that the principal contractor has sole responsibility for the preparation of the construction phase plan. The success in improving the management of health and safety in construction will rely heavily on the effectiveness of the construction phase plan. The 'transparency' of the health and safety plan and its effectiveness will also be one of the measures that the Executive will be able to use as part of monitoring and investigating any accidents. It will undoubtedly be the first source of information that any health and safety inspector would wish to see on visiting a site or investigating an accident. The construction phase plan is likely to be vital evidence in the Executive's decision-making process when considering enforcement action or prosecution. Equally, in pursuance of any civil proceedings for injury or death, the construction phase plan will be important evidence.

The duty to prepare the construction phase plan

The principal contractor's duty in respect of the construction phase plan is set out in regulation 23. Regulation 23 divides the responsibility for the construction phase plan into three distinct phases. These may be identified as the preparation of the construction phase plan, revision of the construction phase plan and finally, implementation.

The first stage, which is preparation of the plan governed by regulation 23(1)(a), is as follows:

> *The principal contractor shall before the start of the construction phase, prepare a construction phase plan which is sufficient to ensure that the construction phase is planned, managed and monitored in a way which enables the construction work to be started so far as is reasonably practicable without risk to health or safety, paying adequate regard to the information provided by the designer under regulations 11(6) and 18(2) and the pre-construction information provided under regulation 20(2)(b).*

Thus the principal contractor has the responsibility to prepare a construction phase plan. The plan has to be sufficient to ensure that the construction phase is planned, managed and monitored. This implies that the construction phase plan should include management procedures for the management and monitoring of the construction work.

The construction plan should also be sufficiently completed so that construction work can be started so far as is reasonably practicable without risk to health and safety. The reference to the information to be provided by the designer under regulations 11(6) and 18(2) implies that there would be no defence to starting construction work if the design information had not been received and incorporated within the construction phase plan. Indeed, the requirements of regulation 23(1)(a) establish the rule that construction work cannot start without the design information having been made available to the principal contractor. It is also a rule that the principal contractor cannot start work until the pre-construction information from the CDM co-ordinator has been received and assimilated.

On large projects, it is not possible for a design to be completed before construction work commences. The ACOP recognises this as a reality, but stresses the need that the construction phase plan for the initial phase of the plan for the construction work must be prepared before any work begins. A construction phase plan should address later activities that are still to be subject to design at least in outline and, for this reason, the monitoring of the plan is an important consideration for complying with regulation 23(1)(b).

To avoid the bureaucratic approach which bedevilled the health and safety plan under the 1994 Regulations, the ACOP stresses that the construction phase plan should not be a collection of generic risk assessments, records of how decisions were reached or detailed method statements. The plan should certainly include the management structure and various parties involved in the project and must be focused on the particular issues arising from that project.

The construction phase plan is a method of communication and should not be confined to long written narratives but, as recognised by the ACOP, can benefit from photographs, sketches, tables and graphs.

The organisation of the construction phase plan should be such that the relevant parts can be easily distributed to other contractors and to the designers.

The principal contractor is required to keep the construction phase plan under review throughout the project. This duty is set out in regulation 23(1)(b) as follows:

> *The principal contractor shall from time to time and as often as may be appropriate throughout the project update, review, revise and refine the construction phase plan so that it continues to be sufficient to ensure that the construction phase plan is planned, managed and monitored in a way which enables the construction work to be carried out so far as is reasonably practicable without risk to health or safety.*

The construction phase plan is therefore a changing and dynamic document. Any change in the design, alternative construction techniques, unusual environmental factors and accidents should all be reasons for a principal contractor to consider the adequacy and effectiveness of the construction phase plan.

The need for revising and monitoring the construction phase plan is essential when the planned construction work has yet to be designed for later phases at the start of construction work. In any event, the construction phase plan should be routinely reviewed, revised and refined by the principal contractor as the construction work proceeds.

The construction phase plan is not intended to be an overly elaborate, obtuse and obscure document. To be effective, the construction phase plan has to be implemented and this is recognised by regulation 23(1)(c) as follows:

> *The principal contractor shall arrange for the construction phase plan to be implemented in a way which will ensure so far as is reasonably practicable the health and safety of all persons carrying out the construction work and all persons who may be affected by the work.*

In preparing the construction phase plan, risks to workers on the construction site, adjacent sites and the public at large should have been considered. Implementation means that the construction phase plan must be communicated to the other parties and duty holders involved in the project. Crucially, the central role of the client should mean that monitoring arrangements proposed by the principal contractor are discussed and agreed with the client.

The distribution of the relevant contents of the construction phase plan have to be the starting-point for implementation. Duty holders and others involved in the project have a requirement to comply with the plan and where the principal contractor recognises that the plan is not being followed must take appropriate action. The principal contractor has the authority to issue reasonable instructions for the purposes of ensuring the health and safety of everyone on site. Equally, anyone receiving such instructions is required to co-operate with complying with such instructions.

There is no requirement on the principal contractor to have the construction phase plan checked by any other duty holder other than obtaining the consent of the client to proceed with construction work. The client is not required to check the details of the plan but he or she must be satisfied that the principal contractor has prepared a construction plan that complies with regulations 23(1)(a) and 23(2).

The client is not required to check any details which the principal contractor proposes for implementing the construction phase plan and its subsequent monitoring and review.

Regulation 23(1) is directed at the purpose and intent of the construction phase plan. It is regulation 23(2) which requires the principal contractor to include in the construction phase plan the measures which address the specific project risks to health and safety. Regulation 23(2) states as follows:

> *The principal contractor shall take all reasonable steps to ensure that the construction phase plan identifies the risk to health and safety arising from the construction work (including the risks specific to the particular type of construction work concerned) and includes suitable and sufficient measures to address such risks, including any site rules.*

The principal contractor's obligations as to what should be included in the construction phase plan may be summarised as:

- identification of risks to health and safety;

- risks specific to the particular type of construction work;

- suitable and sufficient measures to address such risks; and

- the site rules.

For the principal contractor to be confident that the risks to health and safety have been identified requires that all the information provided by the CDM co-ordinator pursuant to regulation 22(b) and sufficient information provided by the designer under regulations 7(6) and 18(2) together with method statements, site rules and measures taken by sub-contractors and other contractors on the site are assessed and incorporated before the construction phase plan can be completed.

There is likely to be for every project a set of risks to health and safety that arise from the local conditions for the site and the method and sequence of construction. The risks which are not generic should be clearly highlighted in the construction phase plan. The principal contractor could do much worse than comparing the risks which have been identified as project specific with the risks in Part 4 of the Regulations as a means of identifying any omissions.

For every risk that has been identified for the project the principal contractor is required to set out suitable and sufficient measures either to eliminate the risk or reduce the risk to an acceptable level. Each contractor should have adopted the same approach for his or her own construction work but the principal contractor should review the construction work to be undertaken by all the contractors to identify any conflicts and take such measures as are necessary to eliminate those particular risks.

The requirement for the site rules to be included in the construction phase plan should leave the principal contractor in no doubt that the readership of the construction phase plan should include all persons engaged on construction work and all others authorised to enter upon the construction site. The principal contractor should make sure that the client, designers and contractors have satisfied themselves that the information they have provided has been included in the construction phase plan and that the implications of that information have been fully understood and implemented. Although it is always the principal contractor's responsibility to prepare the construction phase plan, it has the opportunity of using the construction phase plan as a management tool to encourage co-operation and co-ordination with and between those other parties. This is reinforced by the fact that the construction phase plan should be kept under review and a prudent principal contractor will allow time within the agenda of any site meetings for discussion and review of the construction phase plan.

Checklist for content of construction phase plan

Project particulars	
1.	Name and address of the client.
2.	Name and address of the CDM co-ordinator.
3.	Names and addresses of the designers.
4.	Names and addresses of the contractors.
5.	Description of the project and identification of specific structures.
6.	Project programme with key dates.
7.	Statement of the principal contractor's health and safety policy.
Health and safety management	
8.	Management structure of the principal contractor and all other participants of the management structure and for the project identifying responsibilities generally but, in particular, with regard to health and safety.
9.	Guidelines for monitoring and review of health and safety performance and health and safety targets.
10.	Arrangements for:
	a. meetings and other means to promote co-operation and co-ordination between parties on the site;
	b. consultation with employees and workers;
	c. exchange of design information required by regulations 11(6) and 18(2) and 20(2)(b);
	d. implementation and communication of design changes during the project;
	e. selection and supervision of contractors;
	f. exchange and implementation of health and safety information between contractors;
	g. site security;
	h. site induction training;
	i. specific training;
	j. welfare facilities, as required by Schedule 2 and first aid;
	k. reporting and investigation of accidents and other incidents including near misses;

Project particulars	
	l. production and approval of risk assessments and method statements and systems of work;
	m. site rules;
	n. fire and emergency procedures.

Identification of significant site risks and the elimination or mitigation of such risks

11.	Safety risks including:
	a. delivery and removal of materials, plant and machinery being taken onto and off the site. Particular risks include traffic accidents, spillages, falling of unstable loads, loss of control;
	b. utilities, including supply of water, electricity and gas together with direct works undertaken by utilities. Particular risks include unintentional interference with pipes and cables, including overhead power lines accommodating adjacent land use, including excessive noise, vibration, light glare, escape of noxious materials, traffic movement; energy distribution installations
	c. stability of structures while carrying out construction work with particular regard to temporary support or structures or transient loading conditions;
	d. demolition or dismantling of a structure, or part of a structure;
	e. the storage, handling or use of explosives;
	f. work on excavations and control of unstable ground conditions including groundwater;
	g. cofferdams and caissons;
	h. prevention of drowning where there is work on, over or near water including impounding of water;
	i. work involving diving;
	j. traffic routes and segregation of vehicles and pedestrians;
	k. fire detection and fire fighting;
	l. lighting conditions during day and night-time conditions;
	m. preventing falls when working at height is necessary;

	Project particulars	
	n.	working with, over or near fragile materials;
	o.	control of lifting operations;
	p.	storage of materials and work equipment;
	q.	maintenance of plant and equipment.
12.	Health risks	
	a.	removal of asbestos, excavation and exposure to chemically or bacterialogically contaminated land;
	b.	manual handling;
	c.	use of and exposure to hazardous substances, noise and vibration, with particular regard to the proximity and conditions of the source of the noise and vibration;
	d.	work with ionising radiation;
	e.	exposure to extreme weather conditions including UV radiation from the sun, high winds and very low temperatures;
	f.	any other significant health risks.
	Health and safety file	
13.	Arrangements and standing instructions for the collection and receipt of information from the contractors.	
14.	Collation and storage of information.	
15.	Arrangements for handing over the information to the CDM co-ordinator.	

This checklist has been based upon the list suggested by the ACOP together with additional suggestions.

14 The health and safety file

Definition
When is the health and safety file required?
What information should be included in the health and safety file?
What happens to the health and safety file on completion of the construction work?

Definition

The health and safety file is defined in regulation 2(1) as meaning:

(a) ... the record referred to in Regulation 20(2)(e); and

(b) includes a health and safety file prepared under Regulation 14(d) of the Construction Design and Management Regulations 1994.

The definition recognises that under the 1994 Regulations there will already have been created a health and safety file in respect of many projects and structures. However, the definition of the health and safety file in the 1994 Regulations referred to:

a file or other record in permanent form.

The reference to 'permanent form' is absent from the definition for new health and safety files created under the Regulations. Health and safety files created under the 1994 Regulations, and are subsequently updated, will continue to be in a permanent form.

The need for the health and safety file to be kept by the client for inspection for an indefinite period of time necessarily implies that whatever form the health and safety file is kept in, it should be in a permanent form. The Regulations do not require that the health and safety file should be held in any particular form or format. Thus, the health and safety file can be a paper or electronic record. It should be noted that photocopies, faxes and certain inks are prone to fade and should not therefore be considered as permanent.

The health and safety file should be easy to store securely and should enable easy retrieval of information. This is a requirement, not to be underestimated, for large projects.

When is the health and safety file required?

By virtue of the fact that the definition refers to one of the general duties of CDM co-ordinators, as set out in regulation 20(2)(e), it is apparent that there is no requirement to create a new health and safety file or update a pre-existing health and safety file, unless the construction work is notifiable when assessed by the criteria in regulation 2(3). Conversely, if a CDM co-ordinator is appointed pursuant to regulation 14(1) by the client, the obligation to create a health and safety file arises.

What information should be included in the health and safety file?

New health and safety files created under the Regulations are records which are defined in more detail by regulation 20(2)(e), which describes the contents of the health and safety file as:

> *containing information relating to the project which is likely to be needed during any subsequent construction work to ensure the health and safety of any person, including the information provided in pursuance of regulations 17(1), 18(2) and 22(1)(j).*

The client is the first person to have any duty in relation to the health and safety file which is the obligation to have information provided to the CDM co-ordinator in accordance with regulation 17(1) which states:

> *The client shall ensure that the CDM co-ordinator is provided with all the health and safety information in the client's possession (or which is reasonably obtainable) relating to the project which is likely to be needed for inclusion in the health and safety file, including information specified in 4(9)(c) of the Control of Asbestos at Work Regulations 2006(a).*

The information specified in regulation 4(9)(c) of the Control of Asbestos at Work Regulations 2002, is:

> *The measures to be specified in the plan for managing the risk shall include adequate measures for ensuring that information about the location and condition of any asbestos or any such substances is –*

(i) provided to every person liable to disturb it; and

(ii) made available to the emergency services.

The client will know that it has to provide the information pursuant to regulation 17(1) because the CDM co-ordinator will have provided the required advice and assistance in compliance with regulation 20(1)(a).

The designer has a duty to provide information in pursuance of regulation 18(2) which states:

> *The designer shall take all reasonable steps to provide with his design sufficient information about aspects of the design of the structure or its construction or maintenance as will adequately assist the CDM co-ordinator to comply with his duties under these Regulations, including his duties in relation to the health and safety file.*

The design information should be compiled after the designer has given due consideration to the likelihood that such information will be needed in subsequent construction work. The reference to cleaning work in the definition of construction work in regulation 2(1) is unlikely to be routine regular maintenance insofar as it involves:

> *The use of water or an abrasive at high pressure or the use of corrosive or toxic substances.*

The prudent designer should provide to the CDM co-ordinator all the information which he or she used and created in preparing the design. The categories of information cannot be prescribed by an exhaustive list, however those items and information that will commonly be provided are identified in the checklist below.

The principal contractor's duty to provide information to the CDM co-ordinator is set out in regulation 22(1)(j) which provides that:

> *The principal contractor for a project shall identify to each contractor the information relating to the contractor's activity required by the CDM co-ordinator for inclusion in the health and safety file in pursuance of regulation 20(2)(e) and ensure that such information is promptly provided to the CDM co-ordinator.*

The contractor's obligations are mirrored by obligation 19(2)(d) which provides that every contractor shall promptly provide to the principal contractor the information for inclusion in the health and safety file which has been identified in pursuance of regulation 22(1)(j).

Strangely, the principal contractor is not expressly required to provide information to the CDM co-ordinator other than by virtue of requiring contractors to contribute. Thus, the principal contractor is obligated to provide information for the health and safety file having identified its own activities in respect of which the CDM co-ordinator requires information.

A non-exhaustive checklist of the items of information which should be included by the client, designers and contractors for inclusion in the health and safety file is set out below.

1. A brief description of the work.

2. Historic site data.

3. Ground investigation reports and records, site survey information, pre- and post-construction phase and any residual hazards.

4. Investigation reports and records.

5. Photographic record of essential site elements.

6. Statement of design philosophy, structural principles, loads, calculations and applicable design standards.

7. Drawings and plans used throughout the construction process, including drawings prepared for tender purposes.

8. Record drawings and plans of the completed structure showing where appropriate means of safe access to service voids.

9. Health and safety information about maintenance and cleaning operations.

10. Instructions on the handling and/or operation of equipment together with the relevant maintenance manuals with particular regard to removal or dismantling of installed plant and equipment.

11. The results of proofing or load tests.

12. The commissioning test results.

13. Materials used in the structure identifying, in particular, hazardous materials including data sheets prepared and supplied by suppliers.

14. Identification and specification of in-built safety features, for example emergency and fire-fighting systems and fail-safe devices (see also item 9 above).

15. Nature and location of services, particularly gas and fuel pipelines and electricity cables.

The health and safety file is not intended as a repository for all the information generated by a project but only those matters which will help with the planning of future constructions work. It is not necessary therefore to include the following in the health and safety file:

1. The construction phase plan.

2. Information provided to the contractors at the tender stage.

3. Information provided by tenderers.

4. Construction costing.

5. Contract documentation.

6. Method statements.

7. Records, notes and minutes of meetings.

8. Health and safety statistics.

9. Notices to utilities.

What happens to the health and safety file on completion of the construction work?

At the end of the construction phase when the CDM co-ordinator has been provided with all the information and prepared the health and safety file in accordance with regulation 20(2)(e), he or she is required to pass the health and safety file to the client in accordance with regulation 20(2)(f).

On receipt of the file, the client has the duty, set out in regulation 17(2), as follows:

> *Where a single health and safety file relates to more than one project, site or structure, or where it includes other related information, the client shall ensure that the information relating to each site or structure can be easily identified.*

The form and organisation of the health and safety file, although prepared by the CDM co-ordinator, is ultimately the responsibility of the client, who has to ensure that the information relating to each site and structure can be easily identified. The extent to which some time should be devoted to planning the content and layout of the health and safety file should not be underestimated.

In addition to the duty at regulation 17(2), the client has two distinct and continuing duties following the construction phase. First, the client has to keep the health and safety file for inspection as required by regulation 17(3)(a) as follows:

> *The client shall take reasonable steps to ensure that after the construction phase the information in the health and safety file is kept available for inspection by any person who may need it to comply with the relevant statutory provisions.*

A person who requests the client to produce information within the health and safety file cannot be refused, provided that it is pursuant to the relevant statutory provisions. There is no indication as to the time-frame within which the client should produce the information. However, a period of reasonable notice will be implied depending upon all the circumstances.

The client has the duty under regulation 17(3)(a) until such time as he or she no longer has any interests in the project or structures. This is set out in regulation 17(4) which provides:

> *It shall be sufficient compliance with paragraph 3(a) by a client who disposes of his entire interest in the structure if he delivers the health and safety file to the person who acquires his interest in it and ensures that he is aware of the nature and purpose of the file.*

By this means, the new owner of the structure acquires the obligation under regulation 17(3)(a). A client who disposes of something less than the entire interest should ensure that a copy of the relevant part of the health and safety file is provided to the acquirer: typically, tenants under leases and part purchasers. It is important when disposing or acquiring interests in structures that the appropriate warranties and indemnities are put in place by the respective legal advisers with respect to the completeness and accuracy of the health and safety file.

The client cannot just file the health and safety file on a shelf to gather dust having complied with all the duties and confident it can comply with regulation 17(3)(a). Regulation 17(3)(b) requires a client to keep the health and safety file under review as follows:

> *The client shall take reasonable steps to ensure that after the construction phase the information in the health and safety file is revised as often as may be appropriate to incorporate any relevant and new information.*

The client should recognise that revisions may be needed to the health and safety file notwithstanding the fact that subsequent construction is not notifiable and work has been carried out without the need for a construction phase plan.

15 Contract documentation

Introduction
Clients and election
The professional appointments – duty of care
Terms of engagement of the CDM co-ordinator
Terms of engagement of designers
Contractors' tender documentation

Introduction

There is no need to include in a contract, as between the parties to a project, the precise obligations and duties which are set out fully in the Regulations. The requirements and prohibitions as they affect the roles of contracting parties according to the Regulations apply regardless of any contractual arrangement. However, for the sake of clarity and avoiding confusion there are various matters that should be considered when preparing documentation for projects to which the Regulations apply.

If a party to a project has a contractual obligation to perform the role of a duty holder or comply with other obligations imposed by the Regulations, then failure to so comply becomes a breach of contract for which there is a remedy in damages.

An unplanned benefit of the 1994 Regulations associated with compelling the different parties to communicate with each other, had been to import an improved understanding and clarity of the other party's intentions and expectations. The 1994 Regulations probably did as much for an improvement in contractual relationships on projects as they did for health and safety.

Some of the relevant matters for consideration to be included in the project documentation are dealt with below according to the various roles under the Regulations.

Clients and election

It is not uncommon for any number of persons to be associated with the inception of a project, any of whom could be identified as a client. These persons may include lenders, landowners, developers, joint venture members, future purchasers or tenants. Accordingly, the documentation which will record the agreement between the various persons will range from funding agreements to agreements for a lease. To avoid any doubt as to which person will fulfil the role of client for the purposes of the Regulations, the person chosen as the client should be identified in the relevant documentation. The person who elects to act as client should record his or her agreement to the election in writing in accordance with regulation 8. The person appointed to act as client should consider whether he or she wants an indemnity from the other persons who could have been the client should any claim arise from a breach of the Regulations, including failure to provide information or to co-operate. Conversely, the other persons may wish to have a cause of action against the elected client for any claim flowing from his or her breach of the Regulations. Note that an indemnity cannot provide protection against the risk of a criminal prosecution.

The other parties who agreed to one of their number becoming the client should give some consideration to the consequences of an event, such as insolvency, which would interrupt or disable the elected client from fulfilling his role.

Changing the client during the course of a project is not prohibited, although it should be remembered that unless there is a fresh written agreement in accordance with regulation 8 and all parties to the project have been notified, the last person retains all the duties of the client under the Regulations.

The professional appointments – duty of care

In the case of the CDM co-ordinator and designers they will be obliged to perform their respective duties in accordance with the Regulations to avoid the risk of a criminal prosecution. Therefore, in their relationship with the Executive the obligation to perform as required by the Regulations is absolute. The Executive does not have to demonstrate any loss and there is no possibility of a defence which relies on demonstrating they had performed at least as well as any other person reasonably skilled in their particular field of expertise. However, CDM co-ordinators and designers are not likely to be in a position to give such an absolute warranty of compliance with the Regulations to clients and their appointers respectively.

The professional indemnity insurers will not, in most if not all cases, provide cover for obligations owed to clients by professionals that go beyond the duty of care established in the case of *Bolam* v. *Friern Hospital Management Committee* [1957] 2 All ER 118 which was:

> *Where you get a situation which involves the use of some special skill or competence, then the test as to whether there has been negligence or not is ... the standard of the ordinary skilled man exercising and professing to have that special skill.*

Thus, most clients will have to accept that the contractual obligation owed by CDM co-ordinators or designers will be limited to a duty of care expected of a professional in the same field of expertise as the one in which they are practising. Therefore, the consequences for a CDM co-ordinator or designer arise from two concurrent liabilities, one of which is criminal and the other a contractual duty of care.

A breach of the Regulations may result in a criminal conviction without any consequences arising from a breach of contract if there has been no loss to the client. Conversely, it is possible that a failure to perform the requirements and observe the prohibitions under the Regulations may not lead to a criminal conviction (in the discretion of the Executive), although the client may have suffered a loss for which damages are recoverable in civil proceedings.

At the present time, standard forms of contract, which provide for the contractor to design and build, limit the contractor's liability for design to the duty of care expected of a professional in the relevant discipline. While this duty of care for design is unlikely to be disturbed by the Regulations, it will be a matter for negotiation whether a duty of care should be applied to the role of CDM co-ordinator, in circumstances where this role is undertaken by the principal contractor or contractor.

It is to be anticipated that all collateral warranties, given in the future by professionals and contractors, will reflect the new obligations under the Regulations, in the same terms as the main appointments. Even the CDM co-ordinator may be asked to give collateral warranties to provide comfort for lenders, purchasers or tenants that the health and safety file has been prepared in accordance with the Regulations.

By virtue of the Contract (Third Party Rights) Act 1999, unless the rights of third parties have been excluded expressly in the professional appointments, a person who has suffered harm as a result of the construction work may be able to bring a claim in contract against the professional person who is at fault in having breached the Regulations.

Terms of engagement of the CDM co-ordinator

The client has to appoint a CDM co-ordinator for notifiable projects by virtue of regulation 14(1). A written agreement in accordance with regulation 14(5) incorporating the terms of engagement for the appointment of a CDM co-ordinator will be evidence of the appointment.

The terms of engagement should state expressly that the person is appointed as CDM co-ordinator in accordance with the Regulations. The duties of the CDM co-ordinator set out in the Regulations cannot be varied by contract and are deemed to be incorporated.

In negotiating the terms of engagement of the CDM co-ordinator, some of the matters which might be considered include the following:

1. Warranties as to competence and the allocation of adequate resources.

2. The timing and manner of the provision of information by the client.

3. Identification of arrangements for implementing the co-ordination of health and safety measures.

4. Requests for advice by the client.

5. Identification of arrangements for establishing co-ordination and circulating pre-construction information to the relevant parties.

6. Obligations of co-operation.

7. The preparation, timing and manner of the handing over of the health and safety file.

8. Provisions for terminating the appointment of the CDM co-ordinator with particular regard to handing over.

9. Reporting regime by the CDM co-ordinator of information to the client.

10. Fee arrangements (time charge or percentage basis).

CDM co-ordinators should be alert to the tendency in various standard and bespoke forms of engagement to impose wider obligations than are required by the Regulations.

Terms of engagement of designers

The designer is bound at all times by the relevant duties and obligations contained within the Regulations. Accordingly, there is no requirement to set out in any terms of engagement the requirements of regulation 11 and 18 in respect of notifiable projects with which the designer is obliged to comply.

There are certain matters which, it is advised, should be included in the terms of engagement of a designer. These include:

1. The identity of the CDM co-ordinator and a provision for a subsequent appointment.

2. The identities of any other designers known at the time.

3. The identity of the principal contractor if known at the time, and a provision for a subsequent appointment.

4. Arrangements for co-operation and co-ordination.

5. Details of any notices pursuant to regulation 21 if appropriate.

6. Warranty by the designer as to the competence and the allocation of adequate resources.

Contractors' tender documentation

All documentation sent to contractors and sub-contractors inviting them to submit tenders for projects subject to the Regulations, in addition to the conditions of contract, bills of quantities and specification or employer's requirements (as appropriate), should include the following information:

1. Confirmation, or otherwise, that the project is notifiable.

2. The notice in accordance with regulation 21 including:

    (i)   the identity of the CDM co-ordinator;

    (ii) the identity of the designer(s);

    (iii) the identity of the principal contractor (if appointed), or confirmation that the award of the contract will be the appointment of the principal contractor in accordance with regulation 14(2).

3. Pre-construction information.

The invitation to tender should request the tenderers to submit information to enable the client, or other appointer, to assess competence and the allocation of adequate resources. In particular, it is recommended that the tenderers submit a programme illustrating how time has been allocated for the management of health and safety.

Standard conditions of contract

Since the first edition of this book most, if not all, of the organisations responsible for publishing their own standard conditions of contract now incorporate printed amendments to incorporate the requirements of the Regulations.

16 Criminal and civil liability and enforcement

Health and safety and the public debate

The public's perception of the health and safety management of large companies is probably based upon the notion that, whenever possible, the profit motive overcomes the need to spend enough on health and safety measures. The Southall, Paddington and Hatfield rail disasters, and various other rail crashes, the *Herald of Free Enterprise* disaster, the King's Cross fire and the *Marchioness* sinking on the Thames all received wide press coverage and stimulated a debate that has created a powerful lobby to employ the full force of the criminal law against companies and management executives where death or injury is caused by negligence.

The Labour Government responded to the public's sense of outrage that there should be greater accountability for health and safety, by introducing a Bill in 2006 to create the offence of corporate manslaughter. The message is clear enough: health and safety is a management responsibility and criminal proceedings are a significant factor in reminding employees of the importance attached to carrying out their duties under health and safety laws.

Enforcement

Generally

The Regulations are enforced by the Executive in all cases where construction work is being carried out subject always to the enforcement powers of local authorities allocated to them by the Health and Safety (Enforcing Authority) Regulations 1998 or the Office of Rail Regulation for projects where there are any works associated with railways or tramways.

The activities in respect of which the Executive is the enforcing authority together with guidance provided by the Executive to its own inspectors and local authorities are set out in detail in Appendix 2.

The Executive's inspectors have wide-ranging powers enabling them, for example, to require the production of documents, inspection of and have copies taken of such documents. The full extent of an inspector's powers is set out in section 20 of HASWA 1974. The Executive's enforcement powers, if a statutory provision has been contravened, include the service of an improvement notice (requiring a contravention to be remedied within a certain time period) or a prohibition notice (requiring that the activity in question be stopped until the contravention is remedied).

It may reasonably be assumed that the Executive's inspectors will require to see the construction phase plan as one of the first actions during a visit to a construction site. The inspector will be anxious to see that the management system on site has implemented the intentions developed in the construction phase plan.

During visits to the offices of clients, the inspector may be concerned to see how the client has satisfied him- or herself with regard to the competence of his or her appointees including the CDM co-ordinator, designers and principal contractor, or the decision-making process in permitting construction work to commence. Visits to the offices of CDM co-ordinator and designers may lead, typically, to lines of enquiry which will reveal whether the resources applied to the project are adequate, whether they have undertaken an appropriate risk assessment and whether they are co-operating and contributing to the co-ordination of tasks with other parties.

In respect of fire

The enforcing authority, following a fire on a construction site, will be determined by the Regulatory Reform (Fire Safety) Order 2005. The Executive will be the enforcing authority in accordance with article 25(b)(iv) which confirms that:

For the purposes of this Order, 'enforcing authority' means –

(b) the Health and Safety Executive in relation to –

(iv) any workplace which is or is on a construction site within the meaning of regulation 2(1) of the Construction (Design and Management) Regulations 2007 and to which those regulations apply, other than construction sites referred to in regulation 46 of those Regulations.

The categories of construction sites are set out in regulation 46 as follows:

'1.   Subject to paragraphs 2 and 3 –

(a) in England and Wales the enforcing authority within the meaning of article 25 of the Regulatory Reform (Fire Safety) Order 2005; or

(b) in Scotland the enforcing authority within the meaning of section 61 of the Fire (Scotland) Act 2005;

shall be the enforcing authority in respect of a construction site which is contained within or forms part of, premises which are occupied by persons other than those carrying out the construction work or any activity arising from such work as regards regulations 39 and 40, insofar as those regulations relate to fire and regulation 41.

The allocation of responsibility for enforcement with respect to fire as between the Executive and the Fire and Rescue Authority for the area in which the construction site exists only applies to the extent provided by regulations 46(2) and (3) as follows:

2.   In England and Wales paragraph 1 only applies in respect of premises to which the regulation reform (Fire and Safety) Order 2005 applies.

3.   In Scotland paragraph 1 only applies in respect of premises to which Part 3 of the Fire (Scotland) Act 2005 applies.

Thus, a construction site which is within, or forms part of, premises which are a workplace where persons are working is the responsibility of the Fire and Rescue Authority in the event of a fire and not the Executive. In respect of regulation 39, which deals with emergency procedures, regulation 40 which deals with emergency routes and exits, the enforcing authority is the Fire and Rescue Authority insofar as the procedure's routes and exits relate to fire. The Fire and Rescue Authority is responsible for the enforcement of regulation 41 which deals solely with fire detection and fire fighting.

Criminal proceedings

Despite the powers of enforcement, the criminal law is the principal instrument for securing compliance with the duties imposed by HASWA 1974 and all regulations made under it, including the Regulations. Section 33 of HASWA lists 15 separate offences which include the contravention of any health and safety regulations or any requirement or prohibition imposed under any such regulations. The parties to a project are also liable inter alia to a criminal prosecution if they fail to comply with the requirements of the Executive's inspectors, which might, for example, involve the production of the health and safety plan and file and/or the design documentation. Most importantly, directors, managers and other officers of corporate organisations can be prosecuted personally where it is shown they have consented to, or connived at, the commission of any offence or where an offence has been committed by neglect on their part. The Executive has shown in recent years a greater willingness to prosecute individuals in addition to companies and other organisations. There can be no doubt that the trend of prosecuting individuals will continue.

The Magistrates' Courts have jurisdiction, according to the nature of the offence, to impose fines not exceeding £20 000 and a term of up to 6 months imprisonment, or both. Alternatively, if the offence is sufficiently serious to be tried in the Crown Court on indictment, the maximum fine is unlimited and imprisonment for a term not exceeding two years can be imposed, or both. In extreme cases a fine might be imposed in addition to a term of imprisonment.

Guidelines on sentencing

The Court of Appeal in *R. v. F. Howe and Son (Engineers) Limited* established guidelines on sentencing for health and safety offences. The court stressed that every case would have to be decided on its own facts and it avoided laying down any tariff or rule relating to the size of the fine to turnover or net profit. Nonetheless, in general terms, the following criteria were said to be relevant to sentencing:

(a) How far short of the appropriate standard was the defendant's conduct in failing to reach the reasonably practical test?

(b) Death is to be treated as an aggravating feature of an offence and public disquiet should be reflected in the penalty.

(c) What was the degree of risk and the extent of the danger created?

(d)   What was the extent of the breach/breaches, e.g. an isolated incident or continuous over a period?

(e)   What are the defendant's resources and what effect will a fine have on its business?

(f)   Failure to heed warnings will be a particularly aggravating feature.

(g)   Any financial profit that the defendant has deliberately made from failing to take the necessary health and safety steps, or being prepared to run the risk of not taking them to save money, will be a particularly aggravating feature.

Mitigating factors that a Court will take into account will include:

- prompt admission of liability;

- timely plea of guilty;

- steps taken to remedy deficiencies after the Defendant became aware of them; and

- the defendant's good safety record.

In the later Court of Appeal case of *R. v Friskies Petcare Limited* it was recommended, based on the Howe guidelines, that the prosecution should present to the court any aggravating features in addition to the facts.

Under the Company Directors Disqualification Act 1986, a person can be disqualified from holding the office of director of a company after conviction of an indictable offence connected with the management of a company. This may extend to the management of health and safety matters. This extension of the concept of the management of a company was supported by Viscount Ullswater speaking for the Government in the House of Lords who said:

> *in our view, Section 2 of the Company Directors Disqualification Act 1986 is capable of applying to health and safety matters. That Act provides for the court to make a disqualification order against a person connected with the promotion, formation, management or liquidation of a company. We believe that the potential scope of section 2(1) of that Act is very broad and that 'management' includes the management of health and safety.* (Hansard 28/11/91 Col. 1429)

The courts are likely to use their powers under section 2(1) of the Company Directors Disqualification Act 1986 more readily in a climate of improving standards in corporate governance.

A conviction under HASWA 1974, or for any other crime, by virtue of section 11 of the Civil Evidence Act 1968 is admissible as evidence in any subsequent civil proceedings, subject to such evidence being relevant to any issue in those proceedings. In civil proceedings, discussed below, a conviction constitutes the basic fact of a presumption. The defendant would have the uphill task of persuading the court that the verdict beyond reasonable doubt was wrong.

Failure to comply with the ACOP is not in itself an offence, although it may be taken as proof that a person has contravened the relevant regulation (section 17 – Health and Safety at Work etc. Act 1974). It would, however, be open to the accused to prove that compliance with the Regulations had been achieved in some other way.

Under the 1994 Regulations nearly all the prosecutions were usually connected with accidents or other incidents, particularly where children and public safety are at issue. In other words, the Executive has not been proactive in prosecuting as a deterrent but is being reactive to what has already manifested itself as a failure in the management of health and safety.

Civil liability

In addition to the liability of a criminal prosecution, persons in breach of the Regulations may be liable in civil proceedings in certain circumstances.

Section 47(2) of HASWA 1974 provides:

> *Breach of duty imposed by health and safety Regulations shall, so far as it causes damage, be actionable except insofar as the Regulations provide otherwise.*

In the context of HASWA 1974 and the regulations made further to Article 118A of the Treaty of Rome, which seek improvements in occupational health and safety in an extensive and far-reaching programme of legislation, it is difficult to understand why civil liability is excluded specifically from selected regulations.

Breach of statutory duty

Breach of statutory duty is available as a cause of action in certain circumstances as provided for by regulation 45:

> *Breach of a duty imposed by the preceding provisions of these Regulations, other than those imposed by regulations 9(1)(b), 13(6) and (7), 16,*

22(1)(c) and (1), 25(1), (2) and (4), 26 to 44 and Schedule 2, shall not confer a right of action in any civil proceedings insofar as that duty applies for the protection of a person who is not an employee of the person on whom the duty is placed.

Thus a remedy is available to non-employee claimants for breach of statutory duty for the limited circumstances in the Regulations cited in regulation 45 which can be summarised as follows:

Regulations 9(1)(b), 13(6) and (7), 22(1)(c) and (l)	The duty on the client, contractors and the principal contractor to provide welfare facilities; and to take reasonable steps to prevent access by unauthorised persons to the site during the construction phase.
Regulation 16	The duty on the client to ensure that a construction phase plan has been prepared in accordance with the Regulations and appropriate provision for welfare facilities before the commencement of the construction phase.
Regulations 22(1)(c) and (d), 25(1), (2) and (4), 26 to 44 and Schedule 2	The duties on all the named persons in the Regulations to comply with the specific aspects of health and safety management required by Part 4 and the provision of welfare facilities.

Therefore, an unauthorised person who has obtained access to a construction site before the contractor has taken reasonable steps to deny such access, and who is injured, will be able to sue the contractor for damages arising from the breach of statutory duty. There would not be a right for such a person to bring a claim for statutory duty against the contractor for failing to provide training to his employees pursuant to regulation 13(3). The apparent narrow category of circumstances covered by the Regulations for the benefit of third parties is not likely to cause any substantial injustice. In fact, the scope of Part 4 and the specific duties referred to in regulation 45 will provide a remedy of breach of statutory duty in civil proceedings in many instances for third parties who have suffered harm as a result of the breaches of the particular regulations referred to in regulation 45.

Negligence

The express exclusion of a cause of action based on breach of statutory duty does not exclude a common law claim for negligence. The essential elements of negligence are:

1. A duty owed by the defendant to the claimant.

2. Breach of the duty owed by the defendant.

3. Breach of the duty caused damage.

The duty owed by an employer to its employees was summed up in the leading case of *Wilsons and Clyde Coal Co. Ltd* v *English 1938* as:

> *A duty which rests on the employer and which is personal to the employer, to take reasonable care for the safety of his workmen, whether the employer be an individual, a firm, or a company, and whether or not the employer takes any share in the conduct of the operations.*

The duty owed by employers has been categorised by the courts as a duty to provide:

1. a safe workplace;

2. safe equipment;

3. competent staff and fellow workmen;

4. a safe system of work.

The entire philosophy behind the Regulations is to provide a safer workplace on site, together with a safer system of work, or method of construction, and that the CDM co-ordinator, designers and the principal contractor are required to be competent. There will be overlap with other regulations, particularly the Work Equipment Regulations, in respect of the requirement to provide safe equipment and the Management Regulations which deal with general principles of risk assessment.

Although the Regulations do not confer a right of action in civil proceedings, except as provided for in regulation 45, they are evidence of required practice and failure to follow such practice, such as the appointment of a competent designer, can constitute negligence.

Contributory negligence

Many personal injury claims arising from industrial accidents, whether based on breach of statutory duty or negligence, often have to overcome an allegation of contributory negligence. Contributory negligence is proved when:

(i) the injury of which the claimant complains results from that particular risk to which the negligence of the plaintiff exposed him; and

(ii)   the negligence of the claimant contributed to his injury; and

(iii)   there was fault or negligence on the part of the claimant.

The extent to which an employee has suffered as a result of contributory negligence will depend on the facts in each case. Section 7 of HASWA 1974 provides:

> *It shall be the duty of every employee while at work –*
>
> *(a) to take reasonable care for the health and safety of himself and of other persons who may be affected by his acts or omissions at work; and*
>
> *(b) as regards any duty or requirement imposed on his employer or any other person by or under any of the relevant statutory provisions, to co-operate with him so far as is necessary to enable that duty or requirement to be performed or complied with.*

A discussion on contributory negligence of employees and the statutory duties which they owe to their employers is beyond the scope of this book but is mentioned for the sake of completeness.

17 Transitional provisions

The Regulations came into force on 6 April 2007; however, the impact of the Regulations, as the construction industry adjusts to the new regime, is likely to be considerably less burdensome than when the 1994 Regulations came into force.

The transitional provisions are referred to in regulation 47, which takes account of the fact that all notifiable projects that have commenced before 6 April 2007 will already be subject to the 1994 Regulations. Regulation 47 sets out a number of modifications to the Regulations as presaged in regulation 47(1) which provides:

> *These Regulations shall apply in relation to a project which began before their coming into force, with the following modifications.*

Appointment of the CDM co-ordinator and the principal contractor is the first practical step that a client has to undertake, where a project is notifiable, in accordance with regulation 14(1) and regulation 14(2) respectively. Accordingly, where a notifiable project comes within the meaning of the Regulations the client is required to make such appointments as soon as is practicable as referred to in regulation 47(2) which provides:

> *Subject to paragraph (3), where the time specified in paragraph (1) or (2) of regulation 14 for the appointment of the CDM co-ordinator or the principal contractor occurred before the coming into force of these Regulations, the client shall appoint the CDM co-ordinator or, as the case may be, the principal contractor, as soon as is practicable.*

Regulation 47(2) is expressed to be subject to paragraph (3) because, in a large and significant number of notifiable projects, there will already be an appointed planning supervisor and/or principal contractor. The client is forced to consider the question of competence of the pre-existing appointees once again as provided for in regulation 47(3) as follows:

> *Where a client appoints any planning supervisor or principal contractor already appointed under regulation 6 of the Construction (Design and Management) Regulations 1994(d) (referred to in this Regulation as 'the 1994 Regulations') as the CDM co-ordinator or the principal*

contractor respectively pursuant to paragraph (2), regulation 4(1) shall have effect so that the client shall within twelve months of the coming into force of these Regulations take reasonable steps to ensure that any CDM co-ordinator or principal contractor so appointed is competent within the meaning of regulation 4(2).

The initial reappointment of the planning supervisor as CDM co-ordinator and the principal contractor is a stop-gap measure until within 12 months, i.e. on or before 5 April 2008, the client has satisfied itself that the re-appointees are competent within the meaning of regulation 4(2). The provision causes the client to confront whether there is a real and authentic difference between the 1994 Regulations and the Regulations requirements for competence and, further, whether the assessment above for the 1994 Regulations was adequate to meet the requirements of the Regulations.

The extent of the requirement of competence in regulation 4(2) is very similar to the requirements set out in regulation 8(4) of the 1994 Regulations, which states:

Any reference in this regulation to a person having competence shall extend only to his competence –

(a) to perform any requirement; and

(b) to conduct his undertaking without contravening any prohibition, imposed on him by or under any of the relevant statutory provisions.

The client will have to address the new and modified duties, or 'requirements' to use the language of regulation 4(2)(a) of the Regulations and satisfy itself that the appointees can still be considered competent.

The client is relieved of the bureaucratic task of notifying the existing planning supervisor and principal contractor of their appointments under the Regulations if they have already been properly appointed under regulation 6 of the 1994 Regulations by regulation 47(4) which states:

Any planning supervisor or principal contractor appointed under regulation 6 of the 1994 Regulations shall, in the absence of an express appointment by the client, be treated for the purposes of paragraph (2) as having been appointed as the CDM co-ordinator, or the principal contractor, respectively.

The obligations on the planning supervisors and principal contractors appointed under the 1994 Regulations are dealt with in regulation 47(5) which states:

> *Any person treated as having been appointed as the CDM co-ordinator or the principal contractor pursuant to paragraph (4) shall within twelve months of the coming into force of these Regulations take such steps as are necessary to ensure that he is competent within the meaning of regulation 4(2).*

If a planning supervisor and principal contractor has reason to assess itself as not being competent within the 12 months up to 5 April 2008, it has the opportunity to use the remaining balance of time as a period of grace in which to retrain or reallocate resources to achieve competence.

A client, in satisfying itself that a planning supervisor or principal contractor is competent pursuant to paragraph (3), would be prudent to enquire, in the first instance, as to the outcome of any self-assessment of competence that the appointee should have undertaken in satisfaction of paragraph (5). This would avoid the potential embarrassment of the client satisfying itself that the appointee was competent only to discover at a later date that the appointee did not consider itself competent.

The Regulations do not give a client the opportunity to appoint an agent as was possible under regulation 4 of the 1994 Regulations. However, an agent appointed prior to 6 April 2007 can continue in the role of client and subject to all the duties and obligations of the Regulations, subject always to regulation 4 of the 1994 Regulations. Paragraph (6) of regulation 47 sets out the provisions as follows:

> *(6) Any agent appointed by a client or clients under regulation 4 of the 1994 Regulations before the coming into force of these Regulations may, if requested by the client and if he himself consents, continue to act as the agent of that client and shall be subject to such requirements and prohibitions as are placed by these Regulations on that client, unless or until such time as such appointment is revoked by that client or the project comes to an end, or five years elapse from the coming into force of these Regulations, whichever arises first.*

Subject to having consented to continuing in the role of agent for the client, the agent will be relieved of the full effect of the Regulations by the first of one of three events as follows:

1. the agent's appointment is revoked by a client; or

2. the project comes to an end; or

3. the fifth anniversary of the Regulations coming into force i.e. 5 April 2012.

It is implicit that if the agent does not consent to accepting the revised obligations imposed on the client by virtue of the Regulations, the client will be obliged to take on the role with the duty to comply with the Regulations.

If the planning supervisor has given notice to the Executive in accordance with regulation 7 of the 1994 Regulations, the principal contractor can treat this as an effective notice under regulation 21 of the Regulations for the purposes of its obligations under regulation 19(1)(c), subject always to having received a copy of the notice from the planning supervisor, and regulation 22(1)(k) to display the notice. Regulation 47(7) confirms the situation as follows:

> *(7) Where notice has been given under regulation 7 of the 1994 Regulations, the references in regulations 19(1)(c) and 22(1)(k) to notice under regulation 21 shall be construed as being to notice under that regulation.*

Bibliography

Croner Publications (1994) *Croner's Health & Safety at Work* (updated November 1994). Croner Publications Ltd.

Department of the Environment (1994) Press Release, 25 July.

Health and Safety Executive (HSE) (1988) *Blackspot Construction.* HSE.

Oxford English Dictionary (2006), Oxford University Press; 11 Rev. Edn.

Managing Health and Safety in Construction – Construction (Design and Management) Regulations 2007, Approved Code of Practice, Health and Safety Executive (HSE), 2007.

An Outline Map on Competence, Training and Certification.

Appendix 1 The Construction (Design and Management) Regulations 2007 (© Crown Copyright 2007)

PART 1 INTRODUCTION

Citation and commencement

1. These Regulations may be cited as the Construction (Design and Management) Regulations 2007 and shall come into force on 6th April 2007.

Interpretation

2.—(1) In these Regulations, unless the context otherwise requires –

'business' means a trade, business or other undertaking (whether for profit or not);

'client' means a person who in the course or furtherance of a business –

    (a) seeks or accepts the services of another which may be used in the carrying out of a project for him; or

    (b) carries out a project himself;

'CDM co-ordinator' means the person appointed as the CDM co-ordinator under regulation 14(1);

'construction site' includes any place where construction work is being carried out or to which the workers have access, but does not include a workplace within it which is set aside for purposes other than construction work;

'construction phase' means the period of time starting when construction work in any project starts and ending when construction work in that project is completed;

'construction phase plan' means a document recording the health and safety arrangements, site rules and any special measures for construction work;

'construction work' means the carrying out of any building, civil engineering or engineering construction work and includes –

(a) the construction, alteration, conversion, fitting out, commissioning, renovation, repair, upkeep, redecoration or other maintenance (including cleaning which involves the use of water or an abrasive at high pressure or the use of corrosive or toxic substances), de-commissioning, demolition or dismantling of a structure;

(b) the preparation for an intended structure, including site clearance, exploration, investigation (but not site survey) and excavation, and the clearance or preparation of the site or structure for use or occupation at its conclusion;

(c) the assembly on site of prefabricated elements to form a structure or the disassembly on site of prefabricated elements which, immediately before such disassembly, formed a structure;

(d) the removal of a structure or of any product or waste resulting from demolition or dismantling of a structure or from disassembly of prefabricated elements which immediately before such disassembly formed such a structure; and

(e) the installation, commissioning, maintenance, repair or removal of mechanical, electrical, gas, compressed air, hydraulic, telecommunications, computer or similar services which are normally fixed within or to a structure,

but does not include the exploration for or extraction of mineral resources or activities preparatory thereto carried out at a place where such exploration or extraction is carried out;

'contractor' means any person (including a client, principal contractor or other person referred to in these Regulations) who, in the course or furtherance of a business, carries out or manages construction work;

'design' includes drawings, design details, specification and bill of quantities (including specification of articles or substances) relating to a structure, and calculations prepared for the purpose of a design;

'designer' means any person (including a client, contractor or other person referred to in these Regulations) who in the course or furtherance of a business –

(a) prepares or modifies a design; or

(b) arranges for or instructs any person under his control to do so,

relating to a structure or to a product or mechanical or electrical system intended for a particular structure, and a person is deemed to prepare a design where a design is prepared by a person under his control;

'excavation' includes any earthwork, trench, well, shaft, tunnel or underground working;

'the Executive' means the Health and Safety Executive;

'the general principles of prevention' means the general principles of prevention specified in Schedule 1 to the Management of Health and Safety at Work Regulations 1999[3]

'health and safety file' –

(a) means the record referred to in regulation 20(2)(e); and

(b) includes a health and safety file prepared under regulation 14(d) of the Construction (Design and Management) Regulations 1994[4];

'loading bay' means any facility for loading or unloading;

(a) S.I. 1999/3242, to which there are amendments not relevant to these Regulations.

(b) S.I. 1994/3140, amended by S.I. 2006/557; there are other amending instruments but none is relevant.

'place of work' means any place which is used by any person at work for the purposes of construction work or for the purposes of any activity arising out of or in connection with construction work;

'pre-construction information' means the information described in regulation 10 and, where the project is notifiable, regulation 15.

'principal contractor' means the person appointed as the principal contractor under regulation 14(2);

'project' means a project which includes or is intended to include construction work and includes all planning, design, management or other work involved in a project until the end of the construction phase;

'site rules' means the rules described in regulation 22(1)(d);

'structure' means –

(a) any building, timber, masonry, metal or reinforced concrete structure, railway line or siding, tramway line, dock, harbour, inland navigation, tunnel, shaft, bridge, viaduct, waterworks, reservoir, pipe or pipe-line, cable, aqueduct, sewer, sewage works, gasholder, road, airfield, sea defence works, river works, drainage works, earthworks, lagoon, dam, wall, caisson, mast, tower, pylon, underground tank, earth retaining structure or structure designed to preserve or alter any natural feature, fixed plant and any structure similar to the foregoing; or

(b) any formwork, falsework, scaffold or other structure designed or used to provide support or means of access during construction work,

and any reference to a structure includes a part of a structure.

'traffic route' means a route for pedestrian traffic or for vehicles and includes any doorway, gateway, loading bay or ramp;

'vehicle' includes any mobile work equipment;

'work equipment' means any machinery, appliance, apparatus, tool or installation for use at work (whether exclusively or not);

'workplace' means a workplace within the meaning of regulation 2(1) of the Workplace (Health, Safety and Welfare) Regulations 1992[5] other than a construction site; and

'writing' includes writing which is kept in electronic form and which can be printed.

(2) Any reference in these Regulations to a plan, rules, document, report or copy includes a plan, rules, document, report or copy which is kept in a form –

(a) in which it is capable of being reproduced as a printed copy when required; and

(b) which is secure from loss or unauthorised interference.

(3) For the purposes of these Regulations, a project is notifiable if the construction phase is likely to involve more than –

    (a) 30 days; or

    (b) 500 person days,

of construction work.

Application

3.—(1) These Regulations shall apply –

    (a) in Great Britain; and

    (b) outside Great Britain as sections 1 to 59 and 80 to 82 of the 1974 Act apply by virtue of article 8(1)(a) of the Health and Safety at Work etc. Act 1974 (Application outside Great Britain) Order 2001[6].

(2) Subject to the following paragraphs of this regulation, these Regulations shall apply to and in relation to construction work.

(3) The duties under Part 3 shall apply only where a project –

    (a) is notifiable; and

    (b) is carried out for or on behalf of, or by, a client.

(4) Part 4 shall apply only in relation to a construction site.

(5) Regulations 9(1)(b), 13(7), 22(1)(c), and Schedule 2 shall apply only in relation to persons at work who are carrying out construction work.

PART 2
GENERAL MANAGEMENT DUTIES APPLYING TO CONSTRUCTION PROJECTS

Competence

4.—(1) No person on whom these Regulations place a duty shall –

    (a) appoint or engage a CDM co-ordinator, designer, principal contractor or contractor unless he has taken reasonable steps to ensure that the person to be appointed or engaged is competent;

(b) accept such an appointment or engagement unless he is competent;

(c) arrange for or instruct a worker to carry out or manage design or construction work unless the worker is –

(i) competent, or

(ii) under the supervision of a competent person.

(2) Any reference in this regulation to a person being competent shall extend only to his being competent to –

(a) perform any requirement; and

(b) avoid contravening any prohibition,

imposed on him by or under any of the relevant statutory provisions.

Co-operation

5. —(1) Every person concerned in a project on whom a duty is placed by these Regulations, including paragraph (2), shall –

(a) seek the co-operation of any other person concerned in any project involving construction work at the same or an adjoining site so far as is necessary to enable himself to perform any duty or function under these Regulations; and

(b) co-operate with any other person concerned in any project involving construction work at the same or an adjoining site so far as is necessary to enable that person to perform any duty or function under these Regulations.

(2) Every person concerned in a project who is working under the control of another person shall report to that person anything which he is aware is likely to endanger the health or safety of himself or others.

Co-ordination

6. All persons concerned in a project on whom a duty is placed by these Regulations shall co-ordinate their activities with one another in a manner which ensures, so far as is reasonably practicable, the health and safety of persons –

(a) carrying out the construction work; and

(b) affected by the construction work.

General principles of prevention

7. —(1) Every person on whom a duty is placed by these Regulations in relation to the design, planning and preparation of a project shall take account of the general principles of prevention in the performance of those duties during all the stages of the project.

(2) Every person on whom a duty is placed by these Regulations in relation to the construction phase of a project shall ensure so far as is reasonably practicable that the general principles of prevention are applied in the carrying out of the construction work.

Election by clients

8. Where there is more than one client in relation to a project, if one or more of such clients elect in writing to be treated for the purposes of these Regulations as the only client or clients, no other client who has agreed in writing to such election shall be subject after such election and consent to any duty owed by a client under these Regulations save the duties in regulations 5(1)(b), 10(1), 15 and 17(1) insofar as those duties relate to information in his possession.

Client's duty in relation to arrangements for managing projects

9. —(1) Every client shall take reasonable steps to ensure that the arrangements made for managing the project (including the allocation of sufficient time and other resources) by persons with a duty under these Regulations (including the client himself) are suitable to ensure that –

(a) the construction work can be carried out so far as is reasonably practicable without risk to the health and safety of any person;

(b) the requirements of Schedule 2 are complied with in respect of any person carrying out the construction work; and

(c) any structure designed for use as a workplace has been designed taking account of the provisions of the Workplace (Health, Safety and Welfare) Regulations 1992 which relate to the design of, and materials used in, the structure.

(2) The client shall take reasonable steps to ensure that the arrangements referred to in paragraph (1) are maintained and reviewed throughout the project.

Client's duty in relation to information
10. —(1) Every client shall ensure that

    (a) every person designing the structure; and

    (b) every contractor who has been or may be appointed by the client,

is promptly provided with pre-construction information in accordance with paragraph (2).

  (2) The pre-construction information shall consist of all the information in the client's possession (or which is reasonably obtainable), including –

    (a) any information about or affecting the site or the construction work;

    (b) any information concerning the proposed use of the structure as a workplace;

    (c) the minimum amount of time before the construction phase which will be allowed to the contractors appointed by the client for planning and preparation for construction work; and

    (d) any information in any existing health and safety file,

which is relevant to the person to whom the client provides it for the purposes specified in paragraph (3).

  (3) The purposes referred to in paragraph (2) are –

    (a) to ensure so far as is reasonably practicable the health and safety of persons –

      (i) engaged in the construction work,

      (ii) liable to be affected by the way in which it is carried out, and

      (iii) who will use the structure as a workplace; and

    (b) without prejudice to sub-paragraph (a), to assist the persons to whom information is provided under this regulation –

      (i) to perform their duties under these Regulations, and

      (ii) to determine the resources referred to in regulation 9(1) which they are to allocate for managing the project.

Duties of designers

11.—(1) No designer shall commence work in relation to a project unless any client for the project is aware of his duties under these Regulations.

(2) The duties in paragraphs (3) and (4) shall be performed so far as is reasonably practicable, taking due account of other relevant design considerations.

(3) Every designer shall in preparing or modifying a design which may be used in construction work in Great Britain avoid foreseeable risks to the health and safety of any person –

    (a) carrying out construction work;

    (b) liable to be affected by such construction work;

    (c) cleaning any window or any transparent or translucent wall, ceiling or roof in or on a structure;

    (d) maintaining the permanent fixtures and fittings of a structure; or

    (e) using a structure designed as a workplace.

(4) In discharging the duty in paragraph (3), the designer shall –

    (a) eliminate hazards which may give rise to risks; and

    (b) reduce risks from any remaining hazards,

and in so doing shall give collective measures priority over individual measures.

(5) In designing any structure for use as a workplace the designer shall take account of the provisions of the Workplace (Health, Safety and Welfare) Regulations 1992 which relate to the design of, and materials used in, the structure.

(6) The designer shall take all reasonable steps to provide with his design sufficient information about aspects of the design of the structure or its construction or maintenance as will adequately assist –

    (a) clients;

    (b) other designers; and

    (c) contractors,

to comply with their duties under these Regulations.

Designs prepared or modified outside Great Britain

12. Where a design is prepared or modified outside Great Britain for use in construction work to which these Regulations apply –

    (a) the person who commissions it, if he is established within Great Britain; or

    (b) if that person is not so established, any client for the project,

shall ensure that regulation 11 is complied with.

Duties of contractors

13. —(1) No contractor shall carry out construction work in relation to a project unless any client for the project is aware of his duties under these Regulations.

    (2) Every contractor shall plan, manage and monitor construction work carried out by him or under his control in a way which ensures that, so far as is reasonably practicable, it is carried out without risks to health and safety.

    (3) Every contractor shall ensure that any contractor whom he appoints or engages in his turn in connection with a project is informed of the minimum amount of time which will be allowed to him for planning and preparation before he begins construction work.

    (4) Every contractor shall provide every worker carrying out the construction work under his control with any information and training which he needs for the particular work to be carried out safely and without risk to health, including –

    (a) suitable site induction, where not provided by any principal contractor;

    (b) information on the risks to their health and safety –

       (i) identified by his risk assessment under regulation 3 of the Management of Health and Safety at Work Regulations 1999, or

       (ii) arising out of the conduct by another contractor of his undertaking and of which he is or ought reasonably to be aware;

    (c) the measures which have been identified by the contractor in consequence of the risk assessment as the measures

he needs to take to comply with the requirements and prohibitions imposed upon him by or under the relevant statutory provisions;

(d) any site rules;

(e) the procedures to be followed in the event of serious and imminent danger to such workers; and

(f) the identity of the persons nominated to implement those procedures.

(5) Without prejudice to paragraph (4), every contractor shall in the case of any of his employees provide those employees with any health and safety training which he is required to provide to them in respect of the construction work by virtue of regulation 13(2)(b) of the Management of Health and Safety at Work Regulations 1999.

(6) No contractor shall begin work on a construction site unless reasonable steps have been taken to prevent access by unauthorised persons to that site.

(7) Every contractor shall ensure, so far as is reasonably practicable, that the requirements of Schedule 2 are complied with throughout the construction phase in respect of any person at work who is under his control.

PART 3
ADDITIONAL DUTIES WHERE PROJECT IS NOTIFIABLE

Appointments by the client where a project is notifiable

14. —(1) Where a project is notifiable, the client shall appoint a person ('the CDM co-ordinator') to perform the duties specified in regulations 20 and 21 as soon as is practicable after initial design work or other preparation for construction work has begun.

(2) After appointing a CDM co-ordinator under paragraph (1), the client shall appoint a person ('the principal contractor') to perform the duties specified in regulations 22 to 24 as soon as is practicable after the client knows enough about

the project to be able to select a suitable person for such appointment.

(3) The client shall ensure that appointments under paragraphs (1) and (2) are changed or renewed as necessary to ensure that there is at all times until the end of the construction phase a CDM co-ordinator and principal contractor.

(4) The client shall –

(a) be deemed for the purposes of these Regulations, save paragraphs (1) and (2) and regulations 18(1) and 19(1)(a) to have been appointed as the CDM co-ordinator or principal contractor, or both, for any period for which no person (including himself) has been so appointed; and

(b) accordingly be subject to the duties imposed by regulations 20 and 21 on a CDM co-ordinator or, as the case may be, the duties imposed by regulations 22 to 24 on a principal contractor, or both sets of duties.

(5) Any reference in this regulation to appointment is to appointment in writing.

Client's duty in relation to information where a project is notifiable
15. Where the project is notifiable, the client shall promptly provide the CDM co-ordinator with pre-construction information consisting of –

(a) all the information described in regulation 10(2) to be provided to any person in pursuance of regulation 10(1);

(b) any further information as described in regulation 10(2) in the client's possession (or which is reasonably obtainable) which is relevant to the CDM co-ordinator for the purposes specified in regulation 10(3), including the minimum amount of time before the construction phase which will be allowed to the principal contractor for planning and preparation for construction work.

The client's duty in relation to the start of the construction phase where a project is notifiable
16. Where the project is notifiable, the client shall ensure that the construction phase does not start unless –

   (a) the principal contractor has prepared a construction phase plan which complies with regulations 23(1)(a) and 23(2); and

   (b) he is satisfied that the requirements of regulation 22(1)(c) (provision of welfare facilities) will be complied with during the construction phase.

The client's duty in relation to the health and safety file

17. —(1) The client shall ensure that the CDM co-ordinator is provided with all the health and safety information in the client's possession (or which is reasonably obtainable) relating to the project which is likely to be needed for inclusion in the health and safety file, including information specified in regulation 4(9)(c) of the Control of Asbestos Regulations 2006 [7].

   (2) Where a single health and safety file relates to more than one project, site or structure, or where it includes other related information, the client shall ensure that the information relating to each site or structure can be easily identified.

   (3) The client shall take reasonable steps to ensure that after the construction phase the information in the health and safety file –

   (a) is kept available for inspection by any person who may need it to comply with the relevant statutory provisions; and

   (b) is revised as often as may be appropriate to incorporate any relevant new information.

   (4) It shall be sufficient compliance with paragraph (3)(a) by a client who disposes of his entire interest in the structure if he delivers the health and safety file to the person who acquires his interest in it and ensures that he is aware of the nature and purpose of the file.

Additional duties of designers

18. —(1) Where a project is notifiable, no designer shall commence work (other than initial design work) in relation to the project unless a CDM co-ordinator has been appointed for the project.

   (2) The designer shall take all reasonable steps to provide with his design sufficient information about aspects of the design of the

structure or its construction or maintenance as will adequately assist the CDM co-ordinator to comply with his duties under these Regulations, including his duties in relation to the health and safety file.

Additional duties of contractors

19. —(1) Where a project is notifiable, no contractor shall carry out construction work in relation to the project unless –

(a) he has been provided with the names of the CDM co-ordinator and principal contractor;

(b) he has been given access to such part of the construction phase plan as is relevant to the work to be performed by him, containing sufficient detail in relation to such work; and

(c) notice of the project has been given to the Executive, or as the case may be the Office of Rail Regulation, under regulation 21.

(2) Every contractor shall –

(a) promptly provide the principal contractor with any information (including any relevant part of any risk assessment in his possession or control) which –

(i) might affect the health or safety of any person carrying out the construction work or of any person who may be affected by it,

(ii) might justify a review of the construction phase plan, or

(iii) has been identified for inclusion in the health and safety file in pursuance of regulation 22(1)(j);

(b) promptly identify any contractor whom he appoints or engages in his turn in connection with the project to the principal contractor;

(c) comply with –

(i) any directions of the principal contractor given to him under regulation 22(1)(e), and

(ii) any site rules;

(d) promptly provide the principal contractor with the information in relation to any death, injury, condition or dangerous occurrence which the contractor is required to notify or report under the Reporting of Injuries, Diseases and Dangerous Occurrences Regulations 1995[8].

(3) Every contractor shall –

(a) in complying with his duty under regulation 13(2) take all reasonable steps to ensure that the construction work is carried out in accordance with the construction phase plan;

(b) take appropriate action to ensure health and safety where it is not possible to comply with the construction phase plan in any particular case; and

(c) notify the principal contractor of any significant finding which requires the construction phase plan to be altered or added to.

General duties of CDM co-ordinators
20. —(1) The CDM co-ordinator shall –

(a) give suitable and sufficient advice and assistance to the client on undertaking the measures he needs to take to comply with these Regulations during the project (including, in particular, assisting the client in complying with regulations 9 and 16);

(b) ensure that suitable arrangements are made and implemented for the co-ordination of health and safety measures during planning and preparation for the construction phase, including facilitating –

(i) co-operation and co-ordination between persons concerned in the project in pursuance of regulations 5 and 6, and

(ii) the application of the general principles of prevention in pursuance of regulation 7; and

(c) liaise with the principal contractor regarding –

(i) the contents of the health and safety file,

(ii) the information which the principal contractor needs to prepare the construction phase plan, and

        (iii) any design development which may affect planning and management of the construction work.

(2) Without prejudice to paragraph (1) the CDM co-ordinator shall –

    (a) take all reasonable steps to identify and collect the pre-construction information;

    (b) promptly provide in a convenient form to –

        (i) every person designing the structure, and

        (ii) every contractor who has been or may be appointed by the client (including the principal contractor),

such of the pre-construction information in his possession as is relevant to each;

    (c) take all reasonable steps to ensure that designers comply with their duties under regulations 11 and 18(2);

    (d) take all reasonable steps to ensure co-operation between designers and the principal contractor during the construction phase in relation to any design or change to a design;

    (e) prepare, where none exists, and otherwise review and update a record ('the health and safety file') containing information relating to the project which is likely to be needed during any subsequent construction work to ensure the health and safety of any person, including the information provided in pursuance of regulations 17(1), 18(2) and 22(1)(j); and

    (f) at the end of the construction phase, pass the health and safety file to the client.

Notification of project by the CDM co-ordinator

21. —(1) The CDM co-ordinator shall as soon as is practicable after his appointment ensure that notice is given to the Executive containing such of the particulars specified in Schedule 1 as are available.

(2) Where any particulars specified in Schedule 1 have not been notified under paragraph (1) because a principal contractor has not yet been appointed, notice of such particulars shall

be given to the Executive as soon as is practicable after the appointment of the principal contractor, and in any event before the start of the construction work.

(3) Any notice under paragraph (1) or (2) shall be signed by or on behalf of the client or, if sent by electronic means, shall otherwise show that he has approved it.

(4) Insofar as the project includes construction work of a description for which the Office of Rail Regulation is made the enforcing authority by regulation 3(1) of the Health and Safety (Enforcing Authority for Railways and Other Guided Transport Systems) Regulations 2006[9], paragraphs (1) and (2) shall have effect as if any reference to the Executive were a reference to the Office of Rail Regulation.

Duties of the principal contractor
22. —(1) The principal contractor for a project shall –

(a) plan, manage and monitor the construction phase in a way which ensures that, so far as is reasonably practicable, it is carried out without risks to health or safety, including facilitating –

(i) co-operation and co-ordination between persons concerned in the project in pursuance of regulations 5 and 6, and

(ii) the application of the general principles of prevention in pursuance of regulation 7;

(b) liaise with the CDM co-ordinator in performing his duties in regulation 20(2)(d) during the construction phase in relation to any design or change to a design;

(c) ensure that welfare facilities sufficient to comply with the requirements of Schedule 2 are provided throughout the construction phase;

(d) where necessary for health and safety, draw up rules which are appropriate to the construction site and the activities on it (referred to in these Regulations as 'site rules');

(e) give reasonable directions to any contractor so far as is necessary to enable the principal contractor to comply with his duties under these Regulations;

(f) ensure that every contractor is informed of the minimum amount of time which will be allowed to him for planning and preparation before he begins construction work;

(g) where necessary, consult a contractor before finalising such part of the construction phase plan as is relevant to the work to be performed by him;

(h) ensure that every contractor is given, before he begins construction work and in sufficient time to enable him to prepare properly for that work, access to such part of the construction phase plan as is relevant to the work to be performed by him;

(i) ensure that every contractor is given, before he begins construction work and in sufficient time to enable him to prepare properly for that work, such further information as he needs –

   (i) to comply punctually with the duty under regulation 13(7), and

   (ii) to carry out the work to be performed by him without risk, so far as is reasonably practicable, to the health and safety of any person;

(j) identify to each contractor the information relating to the contractor's activity which is likely to be required by the CDM co-ordinator for inclusion in the health and safety file in pursuance of regulation 20(2)(e) and ensure that such information is promptly provided to the CDM co-ordinator;

(k) ensure that the particulars required to be in the notice given under regulation 21 are displayed in a readable condition in a position where they can be read by any worker engaged in the construction work; and

(l) take reasonable steps to prevent access by unauthorised persons to the construction site.

(2) The principal contractor shall take all reasonable steps to ensure that every worker carrying out the construction work is provided with –

(a) a suitable site induction;

(b) the information and training referred to in regulation 13(4) by a contractor on whom a duty is placed by that regulation; and

(c) any further information and training which he needs for the particular work to be carried out without undue risk to health or safety.

The principal contractor's duty in relation to the construction phase plan

23. —(1) The principal contractor shall –

(a) before the start of the construction phase, prepare a construction phase plan which is sufficient to ensure that the construction phase is planned, managed and monitored in a way which enables the construction work to be started so far as is reasonably practicable without risk to health or safety, paying adequate regard to the information provided by the designer under regulations 11(6) and 18(2) and the pre-construction information provided under regulation 20(2)(b);

(b) from time to time and as often as may be appropriate throughout the project update, review, revise and refine the construction phase plan so that it continues to be sufficient to ensure that the construction phase is planned, managed and monitored in a way which enables the construction work to be carried out so far as is reasonably practicable without risk to health or safety; and

(c) arrange for the construction phase plan to be implemented in a way which will ensure so far as is reasonably practicable the health and safety of all persons carrying out the construction work and all persons who may be affected by the work.

(2) The principal contractor shall take all reasonable steps to ensure that the construction phase plan identifies the risks to health and safety arising from the construction work (including the risks specific to the particular type of construction work concerned) and includes suitable and

sufficient measures to address such risks, including any site rules.

The principal contractor's duty in relation to co-operation and consultation with workers

24. The principal contractor shall –

   (a) make and maintain arrangements which will enable him and the workers engaged in the construction work to co-operate effectively in promoting and developing measures to ensure the health, safety and welfare of the workers and in checking the effectiveness of such measures;

   (b) consult those workers or their representatives in good time on matters connected with the project which may affect their health, safety or welfare, so far as they or their representatives are not so consulted on those matters by any employer of theirs;

   (c) ensure that such workers or their representatives can inspect and take copies of any information which the principal contractor has, or which these Regulations require to be provided to him, which relates to the planning and management of the project, or which otherwise may affect their health, safety or welfare at the site, except any information –

      (i) the disclosure of which would be against the interests of national security,

      (ii) which he could not disclose without contravening a prohibition imposed by or under an enactment,

      (iii) relating specifically to an individual, unless he has consented to its being disclosed,

      (iv) the disclosure of which would, for reasons other than its effect on health, safety or welfare at work, cause substantial injury to his undertaking or, where the information was supplied to him by some other person, to the undertaking of that other person, or

      (v) obtained by him for the purpose of bringing, prosecuting or defending any legal proceedings.

PART 4
DUTIES RELATING TO HEALTH AND SAFETY ON CONSTRUCTION SITES

Application of Regulations 26 to 44

25. —(1) Every contractor carrying out construction work shall comply with the requirements of regulations 26 to 44 insofar as they affect him or any person carrying out construction work under his control or relate to matters within his control.

(2) Every person (other than a contractor carrying out construction work) who controls the way in which any construction work is carried out by a person at work shall comply with the requirements of regulations 26 to 44 insofar as they relate to matters which are within his control.

(3) Every person at work on construction work under the control of another person shall report to that person any defect which he is aware may endanger the health and safety of himself or another person.

(4) Paragraphs (1) and (2) shall not apply to regulation 33, which expressly says on whom the duties in that regulation are imposed.

Safe places of work

26. —(1) There shall, so far as is reasonably practicable, be suitable and sufficient safe access to and egress from every place of work and to and from every other place provided for the use of any person while at work, which access and egress shall be properly maintained.

(2) Every place of work shall, so far as is reasonably practicable, be made and kept safe for, and without risks to health to, any person at work there.

(3) Suitable and sufficient steps shall be taken to ensure, so far as is reasonably practicable, that no person uses access or egress, or gains access to any place, which does not comply with the requirements of paragraph (1) or (2) respectively.

(4) Every place of work shall, so far as is reasonably practicable, have sufficient working space and be so arranged that it is suitable for any person who is working or who is likely to

work there, taking account of any necessary work equipment present.

Good order and site security

27. —(1) Every part of a construction site shall, so far as is reasonably practicable, be kept in good order and every part of a construction site which is used as a place of work shall be kept in a reasonable state of cleanliness.

(2) Where necessary in the interests of health and safety, a construction site shall, so far as is reasonably practicable and in accordance with the level of risk posed, either –

(a) have its perimeter identified by suitable signs and be so arranged that its extent is readily identifiable; or

(b) be fenced off,

or both.

(3) No timber or other material with projecting nails (or similar sharp object) shall –

(a) be used in any work; or

(b) be allowed to remain in any place,

if the nails (or similar sharp object) may be a source of danger to any person.

Stability of structures

28. —(1) All practicable steps shall be taken, where necessary to prevent danger to any person, to ensure that any new or existing structure or any part of such structure which may become unstable or in a temporary state of weakness or instability due to the carrying out of construction work does not collapse.

(2) Any buttress, temporary support or temporary structure must be of such design and so installed and maintained as to withstand any foreseeable loads which may be imposed on it, and must only be used for the purposes for which it is so designed, installed and maintained.

(3) No part of a structure shall be so loaded as to render it unsafe to any person.

Demolition or dismantling

29. —(1) The demolition or dismantling of a structure, or part of a structure, shall be planned and carried out in such a manner as to prevent danger or, where it is not practicable to prevent it, to reduce danger to as low a level as is reasonably practicable.

(2) The arrangements for carrying out such demolition or dismantling shall be recorded in writing before the demolition or dismantling work begins.

Explosives

30. —(1) So far as is reasonably practicable, explosives shall be stored, transported and used safely and securely.

(2) Without prejudice to paragraph (1), an explosive charge shall be used or fired only if suitable and sufficient steps have been taken to ensure that no person is exposed to risk of injury from the explosion or from projected or flying material caused thereby.

Excavations

31. —(1) All practicable steps shall be taken, where necessary to prevent danger to any person, including, where necessary, the provision of supports or battering, to ensure that –

(a) any excavation or part of an excavation does not collapse;

(b) no material from a side or roof of, or adjacent to, any excavation is dislodged or falls; and

(c) no person is buried or trapped in an excavation by material which is dislodged or falls.

(2) Suitable and sufficient steps shall be taken to prevent any person, work equipment, or any accumulation of material from falling into any excavation;

(3) Without prejudice to paragraphs (1) and (2), suitable and sufficient steps shall be taken, where necessary, to prevent any part of an excavation or ground adjacent to it from being overloaded by work equipment or material;

(4) Construction work shall not be carried out in an excavation where any supports or battering have been provided pursuant to paragraph (1) unless –

    (a) the excavation and any work equipment and materials which affect its safety, have been inspected by a competent person –

        (i) at the start of the shift in which the work is to be carried out,

        (ii) after any event likely to have affected the strength or stability of the excavation, and

        (iii) after any material unintentionally falls or is dislodged; and

    (b) the person who carried out the inspection is satisfied that the work can be carried out there safely.

(5) Where the person who carried out the inspection has under regulation 33(1)(a) informed the person on whose behalf the inspection was carried out of any matter about which he is not satisfied, work shall not be carried out in the excavation until the matters have been satisfactorily remedied.

Cofferdams and caissons

32. —(1) Every cofferdam or caisson shall be –

    (a) of suitable design and construction;

    (b) appropriately equipped so that workers can gain shelter or escape if water or materials enter it; and

    (c) properly maintained.

(2) A cofferdam or caisson shall be used to carry out construction work only if –

    (a) the cofferdam or caisson, and any work equipment and materials which affect its safety, have been inspected by a competent person –

        (i) at the start of the shift in which the work is to be carried out, and

        (ii) after any event likely to have affected the strength or stability of the cofferdam or caisson; and

    (b) the person who carried out the inspection is satisfied that the work can be safely carried out there.

(3) Where the person who carried out the inspection has under regulation 33(1)(a) informed the person on whose behalf the

inspection was carried out of any matter about which he is not satisfied, work shall not be carried out in the cofferdam or caisson until the matters have been satisfactorily remedied.

Reports of inspections

33. —(1) Subject to paragraph (5), the person who carries out an inspection under regulation 31 or 32 shall, before the end of the shift within which the inspection is completed –

(a) where he is not satisfied that the construction work can be carried out safely at the place inspected, inform the person for whom the inspection was carried out of any matters about which he is not satisfied; and

(b) prepare a report which shall include the particulars set out in Schedule 3.

(2) A person who prepares a report under paragraph (1) shall, within 24 hours of completing the inspection to which the report relates, provide the report or a copy of it to the person on whose behalf the inspection was carried out.

(3) Where the person owing a duty under paragraph (1) or (2) is an employee or works under the control of another, his employer or, as the case may be, the person under whose control he works shall ensure that he performs the duty.

(4) The person on whose behalf the inspection was carried out shall –

(a) keep the report or a copy of it available for inspection by an inspector appointed under section 19 of the Health and Safety at Work etc. Act 1974 [*10*] –

(i) at the site of the place of work in respect of which the inspection was carried out until that work is completed, and

(ii) after that for 3 months,

and send to the inspector such extracts from or copies of it as the inspector may from time to time require.

(5) Nothing in this regulation shall require as regards an inspection carried out on a place of work for the purposes of regulations 31(4)(a)(i) and 32(2)(a)(i), the preparation of more than one report within a period of 7 days.

Energy distribution installations

34. —(1) Where necessary to prevent danger, energy distribution installations shall be suitably located, checked and clearly indicated.

(2) Where there is a risk from electric power cables –

    (a) they shall be directed away from the area of risk; or

    (b) the power shall be isolated and, where necessary, earthed; or

    (c) if it is not reasonably practicable to comply with paragraph (a) or (b), suitable warning notices and –

        (i) barriers suitable for excluding work equipment which is not needed, or

        (ii) where vehicles need to pass beneath the cables, suspended protections, or

        (iii) in either case, measures providing an equivalent level of safety,

    shall be provided or (in the case of measures) taken.

(3) No construction work which is liable to create a risk to health or safety from an underground service, or from damage to or disturbance of it, shall be carried out unless suitable and sufficient steps (including any steps required by this regulation) have been taken to prevent such risk, so far as is reasonably practicable.

Prevention of drowning

35. —(1) Where in the course of construction work any person is liable to fall into water or other liquid with a risk of drowning, suitable and sufficient steps shall be taken –

    (a) to prevent, so far as is reasonably practicable, such person from so falling;

    (b) to minimise the risk of drowning in the event of such a fall; and

    (c) to ensure that suitable rescue equipment is provided, maintained and, when necessary, used so that such person may be promptly rescued in the event of such a fall.

(2) Suitable and sufficient steps shall be taken to ensure the safe transport of any person conveyed by water to or from any place of work.

(3) Any vessel used to convey any person by water to or from a place of work shall not be overcrowded or overloaded.

Traffic routes

36. —(1) Every construction site shall be organised in such a way that, so far as is reasonably practicable, pedestrians and vehicles can move safely and without risks to health.

(2) Traffic routes shall be suitable for the persons or vehicles using them, sufficient in number, in suitable positions and of sufficient size.

(3) A traffic route shall not satisfy sub-paragraph (2) unless suitable and sufficient steps are taken to ensure that –

(a) pedestrians or vehicles may use it without causing danger to the health or safety of persons near it;

(b) any door or gate for pedestrians which leads onto a traffic route is sufficiently separated from that traffic route to enable pedestrians to see any approaching vehicle or plant from a place of safety;

(c) there is sufficient separation between vehicles and pedestrians to ensure safety or, where this is not reasonably practicable –

(i) there are provided other means for the protection of pedestrians, and

(ii) there are effective arrangements for warning any person liable to be crushed or trapped by any vehicle of its approach;

(d) any loading bay has at least one exit point for the exclusive use of pedestrians; and

(e) where it is unsafe for pedestrians to use a gate intended primarily for vehicles, one or more doors for pedestrians is provided in the immediate vicinity of the gate, is clearly marked and is kept free from obstruction.

(4) Every traffic route shall be –

(a) indicated by suitable signs where necessary for reasons of health or safety;

(b) regularly checked; and

(c) properly maintained.

(5) No vehicle shall be driven on a traffic route unless, so far as is reasonably practicable, that traffic route is free from obstruction and permits sufficient clearance.

Vehicles

37.—(1) Suitable and sufficient steps shall be taken to prevent or control the unintended movement of any vehicle.

(2) Suitable and sufficient steps shall be taken to ensure that, where any person may be endangered by the movement of any vehicle, the person having effective control of the vehicle shall give warning to any person who is liable to be at risk from the movement of the vehicle.

(3) Any vehicle being used for the purposes of construction work shall when being driven, operated or towed –

(a) be driven, operated or towed in such a manner as is safe in the circumstances; and

(b) be loaded in such a way that it can be driven, operated or towed safely.

(4) No person shall ride or be required or permitted to ride on any vehicle being used for the purposes of construction work otherwise than in a safe place thereon provided for that purpose.

(5) No person shall remain or be required or permitted to remain on any vehicle during the loading or unloading of any loose material unless a safe place of work is provided and maintained for such person.

(6) Suitable and sufficient measures shall be taken so as to prevent any vehicle from falling into any excavation or pit, or into water, or overrunning the edge of any embankment or earthwork.

Prevention of risk from fire etc.

38. Suitable and sufficient steps shall be taken to prevent, so far as is reasonably practicable, the risk of injury to any person during the carrying out of construction work arising from –

(a) fire or explosion;

(b) flooding; or

(c) any substance liable to cause asphyxiation.

Emergency procedures

39. —(1) Where necessary in the interests of the health and safety of any person on a construction site, there shall be prepared and, where necessary, implemented suitable and sufficient arrangements for dealing with any foreseeable emergency, which arrangements shall include procedures for any necessary evacuation of the site or any part thereof.

(2) In making arrangements under paragraph (1), account shall be taken of –

(a) the type of work for which the construction site is being used;

(b) the characteristics and size of the construction site and the number and location of places of work on that site;

(c) the work equipment being used;

(d) the number of persons likely to be present on the site at any one time; and

(e) the physical and chemical properties of any substances or materials on or likely to be on the site.

(3) Where arrangements are prepared pursuant to paragraph (1), suitable and sufficient steps shall be taken to ensure that –

(a) every person to whom the arrangements extend is familiar with those arrangements; and

(b) the arrangements are tested by being put into effect at suitable intervals.

Emergency routes and exits

40. —(1) Where necessary in the interests of the health and safety of any person on a construction site, a sufficient number of suitable emergency routes and exits shall be provided to enable any person to reach a place of safety quickly in the event of danger.

(2) An emergency route or exit provided pursuant to paragraph (1) shall lead as directly as possible to an identified safe area.

(3) Any emergency route or exit provided in accordance with paragraph (1), and any traffic route giving access thereto, shall be kept clear and free from obstruction and, where necessary, provided with emergency lighting so that such emergency route or exit may be used at any time.

(4) In making provision under paragraph (1), account shall be taken of the matters in regulation 39(2).

(5) All emergency routes or exits shall be indicated by suitable signs.

Fire detection and fire-fighting
41. —(1) Where necessary in the interests of the health and safety of any person at work on a construction site there shall be provided suitable and sufficient –

(a) fire-fighting equipment; and

(b) fire detection and alarm systems,

which shall be suitably located.

(2) In making provision under paragraph (1), account shall be taken of the matters in regulation 39(2).

(3) Any fire-fighting equipment and any fire detection and alarm system provided under paragraph (1) shall be examined and tested at suitable intervals and properly maintained.

(4) Any fire-fighting equipment which is not designed to come into use automatically shall be easily accessible.

(5) Every person at work on a construction site shall, so far as is reasonably practicable, be instructed in the correct use of any fire-fighting equipment which it may be necessary for him to use.

(6) Where a work activity may give rise to a particular risk of fire, a person shall not carry out such work unless he is suitably instructed.

(7) Fire-fighting equipment shall be indicated by suitable signs.

Fresh air

42. —(1) Suitable and sufficient steps shall be taken to ensure, so far as is reasonably practicable, that every place of work or approach thereto has sufficient fresh or purified air to ensure that the place or approach is safe and without risks to health.

(2) Any plant used for the purpose of complying with paragraph (1) shall, where necessary for reasons of health or safety, include an effective device to give visible or audible warning of any failure of the plant.

Temperature and weather protection

43. —(1) Suitable and sufficient steps shall be taken to ensure, so far as is reasonably practicable, that during working hours the temperature at any place of work indoors is reasonable having regard to the purpose for which that place is used.

(2) Every place of work outdoors shall, where necessary to ensure the health and safety of persons at work there, be so arranged that, so far as is reasonably practicable and having regard to the purpose for which that place is used and any protective clothing or work equipment provided for the use of any person at work there, it provides protection from adverse weather.

Lighting

44. —(1) Every place of work and approach thereto and every traffic route shall be provided with suitable and sufficient lighting, which shall be, so far as is reasonably practicable, by natural light.

(2) The colour of any artificial lighting provided shall not adversely affect or change the perception of any sign or signal provided for the purposes of health and safety.

(3) Without prejudice to paragraph (1), suitable and sufficient secondary lighting shall be provided in any place where there would be a risk to the health or safety of any person in the event of failure of primary artificial lighting.

PART 5
GENERAL

Civil liability

45. Breach of a duty imposed by the preceding provisions of these Regulations, other than those imposed by regulations 9(1)(b), 13(6) and (7), 16, 22(1)(c) and (l), 25(1), (2) and (4), 26 to 44 and Schedule 2, shall not confer a right of action in any civil proceedings insofar as that duty applies for the protection of a person who is not an employee of the person on whom the duty is placed.

Enforcement in respect of fire

46. —(1) Subject to paragraphs (2) and (3) –

> (a) in England and Wales the enforcing authority within the meaning of article 25 of the Regulatory Reform (Fire Safety) Order 2005[*11*]; or
>
> (b) in Scotland the enforcing authority within the meaning of section 61 of the Fire (Scotland) Act 2005[*12*],
>
> shall be the enforcing authority in respect of a construction site which is contained within, or forms part of, premises which are occupied by persons other than those carrying out the construction work or any activity arising from such work as regards regulations 39 and 40, in so far as those regulations relate to fire, and regulation 41.

> (2) In England and Wales paragraph (1) only applies in respect of premises to which the Regulatory Reform (Fire Safety) Order 2005 applies.

> (3) In Scotland paragraph (1) only applies in respect of premises to which Part 3 of the Fire (Scotland) Act 2005 applies [*13*].

Transitional provisions

47. —(1) These Regulations shall apply in relation to a project which began before their coming into force, with the following modifications.

> (2) Subject to paragraph (3), where the time specified in paragraph (1) or (2) of regulation 14 for the appointment of the CDM co-ordinator or the principal contractor occurred before the coming into force of these Regulations, the client shall appoint

the CDM co-ordinator or, as the case may be, the principal contractor, as soon as is practicable.

(3) Where a client appoints any planning supervisor or principal contractor already appointed under regulation 6 of the Construction (Design and Management) Regulations 1994[*14*] (referred to in this regulation as 'the 1994 Regulations') as the CDM co-ordinator or the principal contractor respectively pursuant to paragraph (2), regulation 4(1) shall have effect so that the client shall within twelve months of the coming into force of these Regulations take reasonable steps to ensure that any CDM co-ordinator or principal contractor so appointed is competent within the meaning of regulation 4(2).

(4) Any planning supervisor or principal contractor appointed under regulation 6 of the 1994 Regulations shall, in the absence of an express appointment by the client, be treated for the purposes of paragraph (2) as having been appointed as the CDM co-ordinator, or the principal contractor, respectively.

(5) Any person treated as having been appointed as the CDM co-ordinator or the principal contractor pursuant to paragraph (4) shall within twelve months of the coming into force of these Regulations take such steps as are necessary to ensure that he is competent within the meaning of regulation 4(2).

(6) Any agent appointed by a client under regulation 4 of the 1994 Regulations before the coming into force of these Regulations may, if requested by the client and if he himself consents, continue to act as the agent of that client and shall be subject to such requirements and prohibitions as are placed by these Regulations on that client, unless or until such time as such appointment is revoked by that client, or the project comes to an end, or five years elapse from the coming into force of these Regulations, whichever arises first.

(7) Where notice has been given under regulation 7 of the 1994 Regulations, the references in regulations 19(1)(c) and 22(1)(k) to notice under regulation 21 shall be construed as being to notice under that regulation.

Revocations and amendments
48. —(1) The revocations listed in Schedule 4 shall have effect.

(2) The amendments listed in Schedule 5 shall have effect.

Signed by authority of the Secretary of State for Work and Pensions.

Bill McKenzie

Parliamentary Under Secretary of State,
Department for Work and Pensions

7th February 2007

SCHEDULE 1

Regulation 21(1), (2) and (4)

PARTICULARS TO BE NOTIFIED TO THE EXECUTIVE (or Office of Rail Regulation)

1. Date of forwarding.

2. Exact address of the construction site.

3. The name of the local authority where the site is located.

4. A brief description of the project and the construction work which it includes.

5. Contact details of the client (name, address, telephone number and any e-mail address).

6. Contact details of the CDM co-ordinator (name, address, telephone number and any e-mail address).

7. Contact details of the principal contractor (name, address, telephone number and any e-mail address).

8. Date planned for the start of the construction phase.

9. The time allowed by the client to the principal contractor referred to in regulation 15(b) for planning and preparation for construction work.

10. Planned duration of the construction phase.

11. Estimated maximum number of people at work on the construction site.

12. Planned number of contractors on the construction site.

13. Name and address of any contractor already appointed.

14.   Name and address of any designer already engaged.

15.   A declaration signed by or on behalf of the client that he is aware of his duties under these Regulations.

SCHEDULE 2

Regulations 9(1)(b), 13(7) and 22(1)(c)

WELFARE FACILITIES

Sanitary conveniences
  1.   Suitable and sufficient sanitary conveniences shall be provided or made available at readily accessible places. So far as is reasonably practicable, rooms containing sanitary conveniences shall be adequately ventilated and lit.

  2.   So far as is reasonably practicable, sanitary conveniences and the rooms containing them shall be kept in a clean and orderly condition.

  3.   Separate rooms containing sanitary conveniences shall be provided for men and women, except where and so far as each convenience is in a separate room, the door of which is capable of being secured from the inside.

Washing facilities
  4.   Suitable and sufficient washing facilities, including showers if required by the nature of the work or for health reasons, shall so far as is reasonably practicable be provided or made available at readily accessible places.

  5.   Washing facilities shall be provided –

     (a) in the immediate vicinity of every sanitary convenience, whether or not provided elsewhere; and

     (b) in the vicinity of any changing rooms required by paragraph 14 whether or not provided elsewhere.

  6.   Washing facilities shall include –

     (a) a supply of clean hot and cold, or warm, water (which shall be running water so far as is reasonably practicable);

     (b) soap or other suitable means of cleaning; and

     (c) towels or other suitable means of drying.

7. Rooms containing washing facilities shall be sufficiently ventilated and lit.

8. Washing facilities and the rooms containing them shall be kept in a clean and orderly condition.

9. Subject to paragraph 10 below, separate washing facilities shall be provided for men and women, except where and so far as they are provided in a room the door of which is capable of being secured from inside and the facilities in each such room are intended to be used by only one person at a time.

10. Paragraph 9 above shall not apply to facilities which are provided for washing hands, forearms and face only.

Drinking water

11. An adequate supply of wholesome drinking water shall be provided or made available at readily accessible and suitable places.

12. Every supply of drinking water shall be conspicuously marked by an appropriate sign where necessary for reasons of health and safety.

13. Where a supply of drinking water is provided, there shall also be provided a sufficient number of suitable cups or other drinking vessels unless the supply of drinking water is in a jet from which persons can drink easily.

Changing rooms and lockers

14. —(1) Suitable and sufficient changing rooms shall be provided or made available at readily accessible places if –

   (a) a worker has to wear special clothing for the purposes of his work; and

   (b) he cannot, for reasons of health or propriety, be expected to change elsewhere,

   being separate rooms for, or separate use of rooms by, men and women where necessary for reasons of propriety.

   (2) Changing rooms shall –

   (a) be provided with seating; and

   (b) include, where necessary, facilities to enable a person to dry any such special clothing and his own clothing and personal effects.

(3) Suitable and sufficient facilities shall, where necessary, be provided or made available at readily accessible places to enable persons to lock away –

    (a) any such special clothing which is not taken home;

    (b) their own clothing which is not worn during working hours; and

    (c) their personal effects.

Facilities for rest

15. —(1) Suitable and sufficient rest rooms or rest areas shall be provided or made available at readily accessible places.

(2) Rest rooms and rest areas shall –

    (a) include suitable arrangements to protect non-smokers from discomfort caused by tobacco smoke;

    (b) be equipped with an adequate number of tables and adequate seating with backs for the number of persons at work likely to use them at any one time;

    (c) where necessary, include suitable facilities for any person at work who is a pregnant woman or nursing mother to rest lying down;

    (d) include suitable arrangements to ensure that meals can be prepared and eaten;

    (e) include the means for boiling water; and

    (f) be maintained at an appropriate temperature.

SCHEDULE 3

Regulation 33(1)(b)

PARTICULARS TO BE INCLUDED IN A REPORT OF INSPECTION

1.   Name and address of the person on whose behalf the inspection was carried out.

2.   Location of the place of work inspected.

3.   Description of the place of work or part of that place inspected (including any work equipment and materials).

4. Date and time of the inspection.

5. Details of any matter identified that could give rise to a risk to the health or safety of any person.

6. Details of any action taken as a result of any matter identified in paragraph 5 above.

7. Details of any further action considered necessary.

8. Name and position of the person making the report.

SCHEDULE 4

Regulation 48(1)

REVOCATION OF INSTRUMENTS

Description of instrument	Reference	Extent of revocation
The Construction (General Provisions) Regulations 1961	S.I. 1961/1580	The whole Regulations
The Health and Safety Information for Employees Regulations 1989	S.I. 1989/682	Regulation 8(3) and part III of the Schedule
The Construction (Design and Management) Regulations 1994	S.I. 1994/3140	The whole Regulations
The Construction (Health, Safety and Welfare) Regulations 1996	S.I. 1996/1592	The whole Regulations
The Health and Safety (Enforcing Authority) Regulations 1998	S.I. 1998/494	In Schedule 3, the entries relating to the Construction (Design and Management) Regulations 1994 and to the Construction (Health, Safety and Welfare) Regulations 1996
The Provision and Use of Work Equipment Regulations 1998	S.I. 1998/2306	In Schedule 4, the entry relating to the Construction (Health, Safety and Welfare) Regulations 1996
The Lifting Operations and Lifting Equipment Regulations 1998	S.I. 1998/2307	In Schedule 2, the entry relating to the Construction (Health, Safety and Welfare) Regulations 1996
The Management of Health and Safety at Work Regulations 1999	S.I. 1999/3242	Regulation 27 In Schedule 2, the entry relating to the Construction (Design and Management) Regulations 1994
The Construction (Design and Management) (Amendment) Regulations 2000	S.I. 2000/2380	The whole Regulations

Description of instrument	Reference	Extent of revocation
The Fire and Rescue Services Act 2004 (Consequential Amendments) (England) Order 2004	S.I. 2004/3168	Article 37
The Work at Height Regulations 2005	S.I. 2005/735	In Schedule 8, the entry relating to the Construction (Health, Safety and Welfare) Regulations 1996
The Regulatory Reform (Fire Safety) Order 2005	S.I. 2005/1541	Schedule 3 paragraph 3
The Fire and Rescue Services Act 2004 (Consequential Amendments) (Wales) Order 2005	S.I. 2005/2929	Article 37
The Fire (Scotland) Act 2005 (Consequential Modifications and Amendments) (No.2) Order 2005	S.S.I. 2005/344	Schedule 1 Part 1 paragraph 18
The Fire (Scotland) Act 2005 (Consequential Modifications and Savings) (No.2) Order 2006	S.S.I. 2006/457	Schedule 1 paragraph 4
The Health and Safety (Enforcing Authority for Railways and Other Guided Transport Systems) Regulations 2006	S.I. 2006/557	Schedule paragraph 4

SCHEDULE 5

Regulation 48(2)

AMENDMENTS

Description of instrument	Reference	Extent of amendment
The Factories Act 1961	1961 c.34, as amended by S.I. 1996/1592	In section 176(1) in the definitions 'building operation' and 'work of engineering construction' for '1994' substitute '2007'
The Fire (Scotland) Act 2005	2005 asp5, as amended by S.I. 2005/2060	For the words in section 61(9)(a)(iv) substitute 'which are a workplace which is, or is on, a construction site (as defined in regulation 2(1) of the Construction (Design and Management) Regulations 2007) and to which those Regulations apply (other than a construction site to which regulation 46(1) of those Regulations applies)'

Description of instrument	Reference	Extent of amendment
The Construction (Head Protection) Regulations 1989	S.I. 1989/2209	For the words in regulation 2(1) substitute 'Subject to paragraph (2) of this regulation, these Regulations shall apply to construction work within the meaning of regulation 2(1) of the Construction (Design and Management) Regulations 2007'
The Workplace (Health Safety and Welfare) Regulations 1992	S.I. 1992/3004, as amended by S.I. 1996/1592	For the words in regulation 3(1)(b) substitute 'a workplace which is a construction site within the meaning of the Construction (Design and Management) Regulations 2007, and in which the only activity being undertaken is construction work within the meaning of those regulations, save that – (i) regulations 18 and 25A apply to such a workplace; and (ii) regulations 7(1A), 12, 14, 15, 16, 18, 19 and 26(1) apply to such a workplace which is indoors'
The Work in Compressed Air Regulations 1996	S.I. 1996/1656	In regulation 2(1) for the words 'the 1996 Regulations' means the Construction (Health, Safety and Welfare) Regulations 1996' substitute 'the 2007 Regulations' means the Construction (Design and Management) Regulations 2007'
		In regulation 3(1) for '1994' substitute '2007' and for the words 'is not excluded by regulation 3(2)' substitute 'is carried out in the course of a project which is notifiable within the meaning of regulation 2(3)'
		In regulation 5(3) for '1994' substitute '2007'
		In regulation 13(2)(a) for the words '19, 20 and 25(3) of the 1996 Regulations' substitute '39, 40 and 44(3) of the 2007 Regulations'
		In regulation 13(2)(d) for the words '20(1) of the 1996 Regulations' substitute '39(1) of the 2007 Regulations'
		In regulation 14(1) for the words '21 of the 1996 Regulations' substitute '41 of the 2007 Regulations'
		In regulation 18(a) for the words 'regulation 22 of the 1996 Regulations' substitute 'Schedule 2 of the 2007 Regulations'
The Railway Safety (Miscellaneous Provisions) Regulations 1997	S.I. 1997/553	In regulation 2(1) in the definition 'construction work' for '1994' substitute '2007'

Description of instrument	Reference	Extent of amendment
The Fire Precautions (Workplace) Regulations 1997	S.I. 1997/1840	In regulation 3(5)(d) for the words 'the Construction (Health, Safety and Welfare) Regulations 1996' substitute 'the Construction (Design and Management) Regulations 2007'
The Health and Safety (Enforcing Authority) Regulations 1998	S.I. 1998/494	In regulation 2(1) in the definitions 'construction work' and 'contractor' for '1994' substitute '2007'
		In Schedule 2 for the words in paragraph 4(a)(i) substitute 'the project which includes the work is notifiable within the meaning of regulation 2(3) of the Construction (Design and Management) Regulations 2007; or'
The Provision and Use of Work Equipment Regulations 1998	S.I. 1998/2306	In regulation 6(5)(e) for the words 'regulation 29 of the Construction (Health, Safety and Welfare) Regulations 1996' substitute 'regulations 31(4) or 32(2) of the Construction (Design and Management) Regulations 2007'
The Gas Safety (Installation and Use) Regulations 1998	S.I. 1998/2451	In regulation 2(4)(d) for '1994' substitute '2007'
The Work at Height Regulations 2005	S.I. 2005/735	In regulation 2(1) in the definition 'construction work' for the words 'the Construction (Health, Safety and Welfare) Regulations 1996' substitute 'the Construction (Design and Management) Regulations 2007'
The Regulatory Reform (Fire Safety) Order 2005	S.I. 2005/1541	In article 25(b)(iv) for the words 'the Construction (Health, Safety and Welfare) Regulations 1996' substitute 'the Construction (Design and Management) Regulations 2007' and for '33' substitute '46'
The Health and Safety (Enforcing Authority for Railways and Other Guided Transport Systems) Regulations 2006	S.I. 2006/557	In regulation 2 in the definition 'construction work' for '1994' substitute '2007'
		For the words in regulation 5(2)(a)(i) substitute 'the project which includes that work is notifiable within the meaning of regulation 2(3) of the Construction (Design and Management) Regulations 2007; and'

EXPLANATORY NOTE

(This note is not part of the Regulations)

1. These Regulations revoke and replace the Construction (Design and Management) Regulations 1994 (S.I. 1994/3140) (Parts 2 and 3) and revoke and re-enact, with modifications, the Construction (Health, Safety and Welfare) Regulations 1996 (S.I. 1996/1592) (Part 4). They implement in Great Britain the requirements of Directive 92/57/EEC (OJ No. L245, 26.8.92, p. 6) ('the Directive') on the implementation of minimum safety and health requirements at temporary or mobile construction sites (eighth individual Directive within the meaning of Article 16(1) of Directive 89/391/EEC), except certain requirements which are implemented in the Work at Height Regulations 2005 (S.I. 2005/735). These Regulations do not apply the client's duties in the Directive to persons who act otherwise than in the course or furtherance of a trade, business, or other undertaking (regulation 2(1)). They apply the client's duties to make appointments and to ensure that a safety and health plan is drawn up only to projects that meet the threshold for notification to the Health and Safety Executive (or to the Office of Rail Regulation (regulation 21(4)).

2. Parts 2 and 3 set out duties in respect of the planning, management and monitoring of health, safety and welfare in construction projects and of the co-ordination of the performance of these duties by duty holders. Duties applicable to all projects, including duties of clients, designers and contractors, are set out in Part 2. These include a duty on every person working under the control of another to report anything that he is aware is likely to endanger health or safety (regulation 5(2)).

3. Part 3 imposes additional duties on clients, designers and contractors (regulations 14 to 19) where the project is notifiable, defined as likely to involve more than 30 days or 500 person days of construction work (regulation 2(3)). These include the duty of the client to appoint a CDM co-ordinator and a principal contractor (regulation 14), whose particular duties are then set out (regulations 20 to 24).

4. The changes which Parts 2 and 3 make in comparison with the Construction (Design and Management) Regulations 1994 include the following –

   (a) All duty holders under the Regulations are to co-operate with any other person at work on the same or any adjoining site in enabling one another to perform their duties (regulation 5).

(b) All duty holders under the Regulations are to co-ordinate their activities to ensure so far as is reasonably practicable the health and safety of persons carrying out or affected by the construction work (regulation 6).

(c) All duty holders under the Regulations are to take account of the general principles of prevention in Schedule 1 to the Management of Health and Safety at Work Regulations 1999 (S.I. 1999/3242) in the performance of their duties and in the carrying out of the construction work (regulation 7).

(d) The client is under a duty to take reasonable steps to ensure that arrangements for managing the project that are suitable to ensure that construction work can be carried out so far as is reasonably practicable without risk to health and safety are made and maintained by duty holders (regulation 9).

(e) The threshold for notification of a construction project is now also the point at which duties including the making of appointments by the client and the duties of the persons so appointed arise (regulations 14 to 24).

(f) The former appointment of a planning supervisor is now replaced by that of the CDM co-ordinator with enhanced duties, in particular in relation to assisting the client and to the co-ordination of health and safety measures (regulations 20 and 21).

(g) The former duty of the planning supervisor to prepare a health and safety plan has been replaced by that of the principal contractor to prepare a construction phase plan (regulation 23).

5. Part 4 sets out duties applicable to all contractors or to others controlling the way in which construction work is carried out (regulation 25(1) and (2)) in respect of measures to be taken to ensure specified aspects of health and safety and to prevent danger from a number of specified hazards.

6. Civil liability is now restricted under these Regulations only in respect of the Part 2 and 3 duties, for which there is civil liability only to employees, except in respect of the duties concerning welfare facilities and to prevent access by any unauthorised person, and of the client's duty concerning the construction phase plan, for which liability is unrestricted (regulation 45).

7. A copy of the regulatory impact assessment prepared in respect of these Regulations can be obtained from the Health and Safety

Executive, Economic Advisers Unit, Rose Court, 2 Southwark Bridge, London SE1 9HS. A copy of the transposition note in relation to implementation of the Directive can be obtained from the Health and Safety Executive, International Branch at the same address. Copies of both these documents have been placed in the Library of each House of Parliament.

Notes:

[1] 1974 c.37; sections 11(2), 15(1) and 50(3) were amended by the Employment Protection Act 1975 c.71, Schedule 15, paragraphs 4, 6 and 16(3) respectively.

[2] As regards Scotland, see also section 57(1) of the Scotland Act 1998 (1998 c.46) which provides that, despite the transfer to the Scottish Ministers by virtue of that Act of functions in relation to observing and implementing obligations under Community law, any function of a Minister of the Crown in relation to any matter shall continue to be exercisable by him as regards Scotland for the purposes specified in section 2(2) of the European Communities Act 1972 (1972 c.68).

[3] S.I. 1999/3242, to which there are amendments not relevant to these Regulations.

[4] S.I. 1994/3140, amended by S.I. 2006/557; there are other amending instruments but none is relevant.

[5] S.I. 1992/3004, amended by S.I. 2002/2174 and S.I. 2005/735; there are other amending instruments but none is relevant.

[6] S.I. 2001/2127.

[7] S.I. 2006/2739.

[8] S.I. 1995/3163, to which there are amendments not relevant to these Regulations.

[9] S.I. 2006/557, to which there are amendments not relevant to these Regulations.

[10] 1974 c.37.

[11] S.I. 2005/1541, to which there are amendments not relevant to these Regulations. All functions of the Secretary of State under the Order, so far as exercisable in relation to Wales, were transferred to the National Assembly for Wales by S.I. 2006/1458.

[12] 2005 asp5. Section 61(9) was amended by S.I. 2005/2060 article 2(1) and (4)(a) and (b).

[13] Section 77(1) was amended and 77(1A) inserted by S.I. 2005/2060 article 2(1) and (6)(a) and (b); section 77A was inserted by S.I. 2005/2060 article 2(1) and (7); section 78(2) was amended by S.S.I. 2005/352 regulation 2 and S.I. 2005/2060 article 2(1) and (8)(a); section 78(3) was amended and 78(5A) inserted by S.I. 2005/2060 article 2(1) and (8)(b) and (c).

[14] S.I. 1994/3140 amended by S.I. 2006/557; there are other amending instruments but none is relevant.

Appendix 2　Health and Safety (Enforcing Authority) Regulations 1998

SCHEDULE 1

Regulation 3(1)

MAIN ACTIVITIES WHICH DETERMINE WHETHER LOCAL
AUTHORITIES WILL BE ENFORCING AUTHORITIES

1. The sale of goods, or the storage of goods for retail or wholesale distribution, except –

    (a) at container depots where the main activity is the storage of goods in the course of transit to or from dock premises, an airport or a railway;

    (b) where the main activity is the sale or storage for wholesale distribution of any substance or preparation dangerous for supply;

    (c) where the main activity is the sale or storage of water or sewage or their by-products or natural or town gas;

    and for the purposes of this paragraph where the main activity carried on in premises is the sale and fitting of motor car tyres, exhausts, windscreens or sunroofs the main activity shall be deemed to be the sale of goods.

2. The display or demonstration of goods at an exhibition for the purposes of offer or advertisement for sale.

3. Office activities.

4. Catering services.

5. The provision of permanent or temporary residential accommodation including the provision of a site for caravans or campers.

6. Consumer services provided in a shop except dry cleaning or radio and television repairs, and in this paragraph 'consumer services' means services of a type ordinarily supplied to persons who receive

them otherwise than in the course of a trade, business or other undertaking carried on by them (whether for profit or not).

7.   Cleaning (wet or dry) in coin operated units in launderettes and similar premises.

8.   The use of a bath, sauna or solarium, massaging, hair transplanting, skin piercing, manicuring or other cosmetic services and thera-peutic treatments, except where they are carried out under the supervision or control of a registered medical practitioner, a dentist registered under the Dentists Act 1984 [*19*], a physiotherapist, an osteopath or a chiropractor.

9.   The practice or presentation of the arts, sports, games, entertain-ment or other cultural or recreational activities except where the main activity is the exhibition of a cave to the public.

10.   The hiring out of pleasure craft for use on inland waters.

11.   The care, treatment, accommodation or exhibition of animals, birds or other creatures, except where the main activity is horse breeding or horse training at a stable, or is an agricultural activity or veterinary surgery.

12.   The activities of an undertaker, except where the main activity is embalming or the making of coffins.

13.   Church worship or religious meetings.

14.   The provision of car parking facilities within the perimeter of an airport.

15.   The provision of child care, or playgroup or nursery facilities.

SCHEDULE 2

Regulation 4(4)(b)

ACTIVITIES IN RESPECT OF WHICH THE HEALTH AND SAFETY EXECUTIVE IS THE ENFORCING AUTHORITY

4.   The following activities carried on at any premises by persons who do not normally work in the premises –

(a) construction work if –

(i)   the project which includes the work is notifiable within the meaning of regulation 2(3) of the Construction (Design and Management) Regulations 2007; or

(ii) the whole or part of the work contracted to be undertaken by the contractor at the premises is to the external fabric or other external part of a building or structure; or

(iii) it is carried out in a physically segregated area of the premises, the activities normally carried out in that area have been suspended for the purpose of enabling the construction work to be carried out, the contractor has authority to exclude from that area persons who are not attending in connection with the carrying out of the work and the work is not the maintenance of insulation on pipes, boilers or other parts of heating or water systems or its removal from them.

The HSE operational circular OC124/11, Health and Safety (Enforcing Authority) Regulations 1998: A–Z guide to allocation provides the following guidance on the reference to 'fabric' in 4(a)(ii) as follows:

> *Fabric is the basic structure, the walls, roof and floor. 'Other external part' refers to something other than the fabric. An external door, for example, is a fixture believed to fall within this description. A door also forms part of the structure, and its construction or repair etc. falls within the definition of 'construction work.*

The HSE guidance clarifies those activities which come within the definition of construction work but where the local authorities have responsibility for enforcement as follows:

> *Non-notifiable construction work which is entirely internal to the building and which is not separated off from the normal operations of the premises where the local authority is the enforcing authority for that class of premises.*

> *Where the only work carried out in a segregated area of local authority enforced premises is the removal or maintenance of insulation on heating or water systems.*

In general, the erection and dismantling of temporary stages, grandstands and other temporary platform arrangements used by the entertainment industry are not construction operations. The specific requirements of the various regulations made under HASW 1974 will not apply to this kind of work. Similarly, the Regulations do not apply to these kinds of structure.

Appendix 3   Pre-construction information

When drawing up the pre-construction information, each of the following topics should be considered. Information should be included where the topic is relevant to the work proposed. The pre-construction information provides information for those bidding for or planning work, and for the development of the construction phase plan. **The level of detail in the information should be proportionate to the risks involved in the project.**

Pre-construction information

1   *Description of project*

    (a) project description and programme details including:

        (i)   key dates (including planned start and finish of the construction phase), and

        (ii)   the minimum time to be allowed between appointment of the principal contractor and instruction to commence work on site;

    (b) details of client, designers, CDM co-ordinator and other consultants;

    (c) whether or not the structure will be used as a workplace (in which case, the finished design will need to take account of the relevant requirements of the Workplace (Health, Safety and Welfare) Regulations 1992);

    (d) extent and location of existing records and plans.

2   *Client's considerations and management requirements*

    (a) arrangements for:

        (i)   planning for and managing the construction work, including any health and safety goals for the project,

(ii)   communication and liaison between client and others,

(iii)   security of the site,

(iv)   welfare provisions;

(b) requirements relating to the health and safety of the client's employees or customers or those involved in the project such as:

    (i)   site hoarding requirements,

    (ii)   site transport arrangements or vehicle movement restrictions,

    (iii)   client permit-to-work systems,

    (iv)   fire precautions,

    (v)   emergency procedures and means of escape,

    (vi)   'no-go' areas or other authorisation requirements for those involved in the project,

    (vii) any areas the client has designated as confined spaces,

    (viii) smoking and parking restrictions.

3   *Environmental restrictions and existing on-site risks*

(a) Safety hazards, including:

    (i)   boundaries and access, including temporary access – for example narrow streets, lack of parking, turning or storage space,

    (ii)   any restrictions on deliveries or waste collection or storage,

    (iii)   adjacent land uses – for example schools, railway lines or busy roads,

    (iv)   existing storage of hazardous materials,

    (v)   location of existing services, particularly those that are concealed – water, electricity, gas, etc,

    (vi)   ground conditions, underground structures or water courses where this might affect the safe use of plant, for example cranes, or the safety of groundworks,

    (vii) information about existing structures – stability, structural form, fragile or hazardous materials, anchorage points for fall arrest systems (particularly where demolition is involved),

(viii)   previous structural modifications, including weakening or strengthening of the structure (particularly where demolition is involved),

(ix)   fire damage, ground shrinkage, movement or poor maintenance which may have adversely affected the structure,

(x)   any difficulties relating to plant and equipment in the premises, such as overhead gantries whose height restricts access,

(xi)   health and safety information contained in earlier design, construction or 'as-built' – drawings, such as details of pre-stressed or post-tensioned structures;

(b)   health hazards, including:

(i)   asbestos, including results of surveys (particularly where demolition is involved),

(ii)   existing storage of hazardous materials,

(iii)   contaminated land, including results of surveys,

(iv)   existing structures containing hazardous materials,

(v)   health risks arising from client's activities.

4   *Significant design and construction hazards*

(a)   significant design assumptions and suggested work methods, sequences or other control measures;

(b)   arrangements for co-ordination of ongoing design work and handling design changes;

(c)   information on significant risks identified during design;

(d)   materials requiring particular precautions.

5   *The health and safety file*

Description of its format and any conditions relating to its content.

Appendix 4   ACOP Appendix 4

Appendix 4 Competence

Core criteria for demonstration of competence:
Companies, contractors, CDM co-ordinators and designers

You need to meet the standards set out in the core criteria table below. **Column 1** of the table lists the elements which should be assessed when establishing whether or not a company is competent for the work which it will be expected to do. **Column 2** lists the standards against which the assessment should be made. **Column 3** gives some examples of how a company might demonstrate that it meets these standards.

Companies do not have to produce all of the evidence listed in column 3 to satisfy the standard – they simply need to produce enough evidence to show that they meet the standard in column 2, taking account of the nature of the project and the risks which the work entails. This requires you to make a judgement as to whether the evidence provided meets the standard to be achieved. **If your judgement is reasonable, and clearly based on the evidence you have asked for and been provided with, you will not be criticised if the company you appoint subsequently proves not to be competent when carrying out the work.**

Remember that assessments should focus on the needs of the particular job and should be proportionate to the risks arising from the work. Unnecessary bureaucracy associated with competency assessment obscures the real issues and diverts effort away from them.

If you employ less than 5 persons you do not have to write down your policy, organisation or arrangements under criteria 1 and 2. However, you do need to demonstrate that your policy and arrangements are adequate in relation to the type of work you do. Assessments of competence will be made easier if your procedures are clear and accessible.

'Contractor', 'Designer' and 'CDM co-ordinator' relate to your function, not to the type of organisation.

	Criteria	Standard to be achieved	Examples of the evidence that you could use to demonstrate you meet the required standard
	Stage 1 assessment		
1	Health and Safety Policy and Organisation for Health and Safety	You are expected to have and implement an appropriate policy, regularly reviewed, and signed off by the Managing Director or equivalent. The policy must be relevant to the nature and scale of your work and set out the responsibilities for health and safety management at all levels within the organisation.	A signed, current copy of the company policy (indicating when it was last reviewed and by whose authority it is published). Guidance on writing company policies for health and safety can be found in HSE free leaflet INDG 259.
2	Arrangements	These should set out the arrangements for health and safety management within the organisation and should be relevant to the nature and scale of your work. They should set out how the company will discharge their duties under CDM 2007. There should be a clear indication of how these arrangements are communicated to the workforce.	A clear explanation of the arrangements which the company has made for putting its policy into effect and for discharging its duties under CDM 2007. Guidance on making arrangements for the management of health and safety can be found in HSE free leaflet INDG 259.
3	Competent Advice – Corporate and Construction related	Your organisation, and your employees, must have ready access to competent health and safety advice, preferably from within your own organisation. The advisor must be able to provide general health and safety advice, and also (from the same source or elsewhere) advice relating to construction health and safety issues.	Name and competency details of the source of advice, e.g. a safety group, trade federation, or consultant who provides health and safety information and advice. An example from the last 12 months of advice given and action taken.

	Criteria	Standard to be achieved	Examples of the evidence that you could use to demonstrate you meet the required standard
4	Training and Information	You should have in place, and implement, training arrangements to ensure your employees have the skills and understanding necessary to discharge their duties as Contractors, Designers or CDM Co-ordinators. You should have in place a programme for refresher training, e.g. a Continuing Professional Development (CDP) programme or life-long learning which will keep your employees updated on new developments and changes to legislation or good health and safety practice. This applies throughout the organisation from Board or equivalent, to trainees.	Headline training records. Evidence of a H&S training culture including records, certificates of attendance and adequate H&S induction training for site-based workforce. Evidence of an active CPD programme. Sample 'tool box talks'.
5	Individual Qualifications and Experience	Employees are expected to have the appropriate qualifications and experience for the assigned tasks, unless they are under controlled and competent supervision.	Details of qualifications and/or experience of specific corporate post holders e.g. Board members, Health and Safety Advisor etc. Other key roles should be named or identified and details of relevant qualifications and experience provided. **FOR CONTRACTORS:** details of number/percentage of people engaged in the project who have passed a construction health and safety assessment, e.g. the CITB Construction Skills touch screen test or affiliated schemes, such as the CCNSG equivalent; For site managers, details of any specific training such as the Construction Skills CITB 'Site Management Safety Training Scheme' certificate or equivalent; For professionals, details of qualifications and/or professional institution membership;

	Criteria	Standard to be achieved	Examples of the evidence that you could use to demonstrate you meet the required standard
			Details of any relevant qualifications or training such as S/NVQ certificates;
			Evidence of a company based training programme suitable for the work to be carried out.
			FOR DESIGN ORGANISATIONS: details of number/percentage of people engaged in the project who have passed a construction health and safety assessment, e.g. the CITB Construction Skills touch screen test or affiliated schemes, or the CCNSG equivalent;
			Details of any relevant qualifications and/or professional Institution membership and Any other specific qualifications such as ICE construction H&S Register, NEBOSH Construction Certificate, APS Design Register;
			FOR CDM CO-ORDINATORS: details of number/percentage of people engaged in the project who have passed a construction health and safety assessment, e.g. the CITB Construction Skills touch screen test or affiliated schemes, or the CCNSG equivalent;
			Evidence of health and safety knowledge such as NEBOSH Construction Certificate;
			Details of Professional Institution membership and any other specific qualifications such as member of the co-ordinators register administered by the APS, ICE construction H&S register etc.
			Evidence of a clear commitment to training and the Continuing Professional Development of staff.
6	Monitoring, Audit and Review	You should have a system for monitoring your procedures, for auditing them at periodic intervals, and for reviewing them on an on-going basis.	Could be through formal audit or discussions/reports to senior managers;
			Evidence of recent monitoring and management response;
			Copies of site inspection reports.

	Criteria	Standard to be achieved	Examples of the evidence that you could use to demonstrate you meet the required standard
7	Workforce involvement	You should have, **and** implement, an established means of consulting with your workforce on health and safety matters.	Evidence showing how consultation is carried out; Records of HS Meetings/Committees; Names of appointed safety representatives (trades union or other); For those employing <5, be able to describe how you consult with your employees to achieve the consultation required.
8	Accident reporting and enforcement action; follow up investigation	You should have records of all RIDDOR reportable events for at least the last 3 years. You should also have in place a system for reviewing all incidents, and recording the action taken as a result. You should record any enforcement action taken against your company over the last 5 years, and the action which you have taken to remedy matters subject to enforcement action.	Evidence showing the way in which you record and investigate accidents and incidents; Records of last 2 accidents/incidents and action taken to prevent recurrence; Records of any enforcement action taken over the last 5 years, and what action was taken to put matters right; (Information on enforcement taken by HSE over the last 5 years is available on the HSE website) For larger companies, simple statistics showing incidence rates of major injuries, over 3-day injuries, reportable cases of ill-health and dangerous occurrences for the last 3 years; Records should include any incidents that occurred whilst the company traded under a different name, and any incidents that occur to direct employees or labour only sub-contractors.
9	Sub-contracting/consulting procedures (if applicable)	You should have arrangements in place for appointing competent sub-contractors/consultants. You should be able to demonstrate how you ensure that sub-contractors will also have arrangements for appointing competent sub-contractors or consultants. You should have arrangements for monitoring sub-contractor performance.	Evidence showing how you ensure sub-contractors are competent; Examples of sub-contractor assessments you have carried out; Evidence showing how you require similar standards of competence assessment from sub-contractors; Evidence showing how you monitor sub-contractor performance.

	Criteria	Standard to be achieved	Examples of the evidence that you could use to demonstrate you meet the required standard
10	Hazard elimination and risk control (**Designers only**)	You should have, and implement, arrangements for meeting your duties under regulation 11 of CDM2007.	Evidence showing how you: Ensure co-operation and co-ordination of design work within the design team and with other designers/contractors. Ensure that hazards are eliminated and any remaining risks controlled. Ensure that any structure which will be used as a workplace will meet relevant requirements of the Workplace (Health Safety and Welfare) Regulations 1992. Examples showing how risk was reduced through design. A short summary of how changes to designs will be managed. (Note: the emphasis here should be on practical measures which reduce particular risks arising from the design, not on lengthy procedural documentation highlighting generic risks.)
11	Risk assessment leading to a safe method of work (**Contractors only**)	You should have procedures in place for carrying out risk assessments and for developing and implementing safe systems of work/method statements.	Evidence showing how the company will identify significant HS risks and how they will be controlled. Sample risk assessments/safe systems of work/method statements. If you employ less than 5 persons and do not have written arrangements, you should be able to describe how you achieve the above.
		The identification of health issues is expected to feature prominently in this system.	This will depend upon the nature of the work, but must reflect the importance of this risk area.
12	Co-operating with others and co-ordinating your work with that of other contractors (**Contractors**)	You should be able to illustrate how co-operation and co-ordination of your work is achieved in practice, and how you involve the workforce in drawing up method statements/safe systems of work.	Evidence could include for sample risk assessments, procedural arrangements, project team meeting notes. Evidence of how the company co-ordinates its work with other trades.

	Criteria	Standard to be achieved	Examples of the evidence that you could use to demonstrate you meet the required standard
13	Welfare Provision **(Contractors)**	You should be able to demonstrate how you will ensure that appropriate welfare facilities will be in place before people start work on site.	Evidence could include for example health and safety policy commitment; contracts with welfare facility providers; details of type of welfare facilities provided on previous projects.
14	CDM Co-ordinator's duties **(CDM Co-ordinators)**	You should be able to demonstrate how you go about encouraging co-operation, co-ordination and communication between designers.	The evidence should be in the form of actual examples rather than by generic procedures.
Stage 2 assessment			
1	Work experience	You should give details of relevant experience in the field of work for which you are applying.	A simple record of recent projects/contracts should be kept, with the phone numbers/addresses of contacts who can verify that work was carried out with due regard to health and safety. This should be sufficient to demonstrate your ability to deal with the key health and safety issues arising from the work you are applying for. Where there are significant shortfalls in your previous experience, or there are risks associated with the project which you have not managed before, an explanation of how these shortcomings will be overcome.

Appendix 5   ACOP Appendix 5

Appendix 5 Competence

Guidance for assessing competence of a co-ordinator for a larger or more complex project, or one with high or unusual risks

Organisations do not have to produce all of the evidence listed in column 4 to satisfy the standard – they simply need to produce enough evidence to show that they meet the standard in column 2, taking account of the nature of the project and the risks which the work entails. This requires you to make a judgement as to whether the evidence provided meets the standard to be achieved. **If your judgement is reasonable, and clearly based on the evidence provided, you will not be criticised if the company you appoint subsequently proves not to be competent to carry out the work.**

Remember that assessments should focus on the needs of the particular job and should be proportionate to the risks arising from the work. **Unnecessary bureaucracy associated with competency assessment obscures the real issues and diverts effort away from them.**

Stage	Knowledge and experience standard	Field of knowledge and experience	Examples of attainment which should indicate competence
	Task knowledge appropriate for the tasks to be undertaken. May be technical or managerial.	The design and construction process.	Professionally Qualified to Chartered level (*Note 1*). Membership of a relevant construction institution, for example CIBSE; ICE; IEE; IMechE; IStructE; RIBA; CIAT; CIOB.
Stage 1	Health and safety knowledge sufficient to perform the task safely, by identifying hazard and evaluating the risk in order to protect self and others, and to appreciate general background.	Health and safety in construction.	Validated CPD in this field, and typical additional qualification, e.g. NEBOSH Construction Certificate, Member of H&S Register administered by the ICE (*Note 2*). Membership of association for project safety; membership of Institution of Construction Safety (formerly the Institution of Planning Supervisors).

Stage	Knowledge and experience standard	Field of knowledge and experience	Examples of attainment which should indicate competence
Stage 2	Experience and ability sufficient to perform the task (including where appropriate an appreciation of constructability), to recognise personal limitations, task related faults and errors and to identify appropriate actions.	Experience relevant to the task.	Evidence of significant work on similar projects with comparable hazards, complexity and procurement route.

Note 1 Chartered membership of a recognised construction related institution.
Note 2 Open to any member of a construction-related institution.

Index